Thomas Schulenberg

Annual Report 2015 of the Institute for Nuclear and Energy Technologies

Karlsruhe Institute of Technology
KIT SCIENTIFIC REPORTS 7720

Annual Report 2015 of the Institute for Nuclear and Energy Technologies

by
Thomas Schulenberg

KIT Scientific Publishing

Report-Nr. KIT-SR 7720

Impressum

KIT Scientific Publishing

Karlsruher Institut für Technologie (KIT)
KIT Scientific Publishing
Straße am Forum 2
D-76131 Karlsruhe

KIT Scientific Publishing is a registered trademark of Karlsruhe Institute of Technology. Reprint using the book cover is not allowed.

www.ksp.kit.edu

Print on Demand 2016

ISSN 1869-9669
ISBN 978-3-7315-0572-3
DOI: 10.5445/KSP/1000057984

Content

Structure and Activities of the Institute for Nuclear and Energy Technologies

Thomas Schulenberg

Mission

The Institute for Nuclear and Energy Technologies (Institut für Kern- und Energietechnik, IKET) is situated with its offices and research laboratories on the North Campus of KIT. It is focused on nuclear, fusion and renewable energy technologies for electric power production and on hydrogen technologies for energy storage in chemical form. Its research topics include analyses and tests of thermal-hydraulic phenomena, combustion phenomena and neutron physics which are typical for normal operation or for accidental conditions in nuclear power plants, for future nuclear fusions reactors, for geothermal or solar power plants, but also for mobile systems. Most subjects are application oriented, supported by some basic research projects, if needed.

Organizational structure

IKET is structured into nine working groups as indicated in Fig. 1. Working groups on accident analyses, on accidents management systems and on transmutation as well as the AREVA Nuclear Professional School have been concentrating in 2015 primarily on nuclear applications, whereas the Karlsruhe Liquid Metal Laboratory (KALLA) and the Hydrogen group were addressing nuclear as well as other energy technologies, as will be outlined below. The working group on magneto-hydrodynamics is primarily working on nuclear fusion applications, whereas the working group on energy and process engineering is rather concentrating on geothermal energies. Thus, the institute covers a wide field of different energy technologies, and the share of its personnel resources on the different research topics is determined each year by the worldwide market request for energy research.

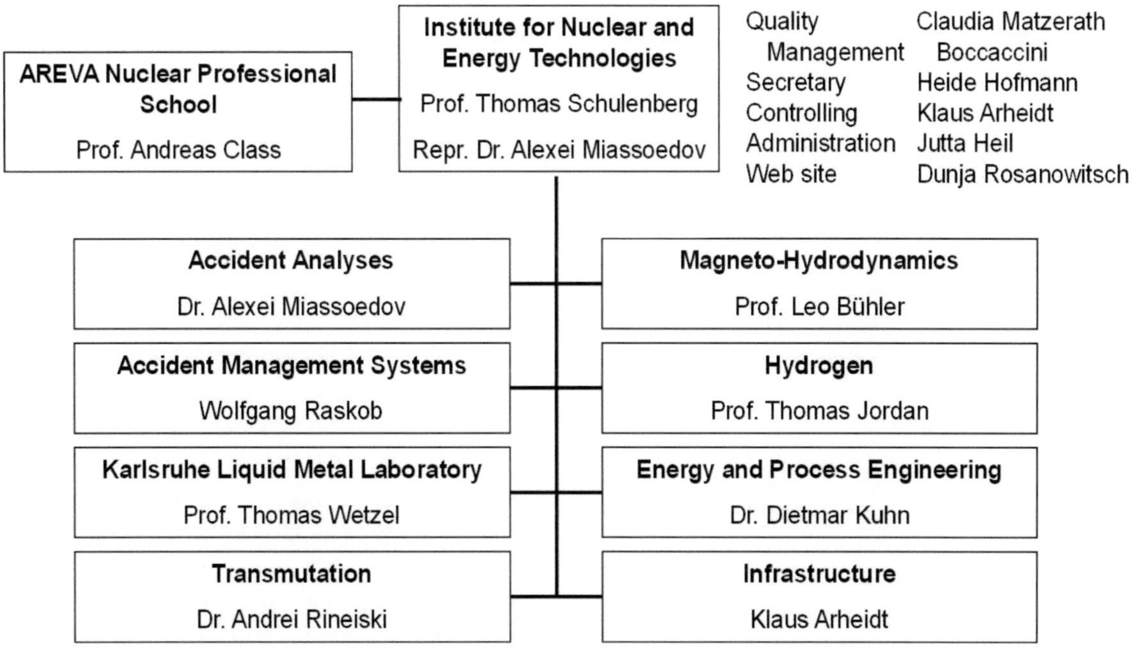

Fig. 1: Organizational structure of the Institute for Nuclear and Energy Technologies (IKET)

Having a larger share of nuclear and fusion research, IKET has a quality management system according to the international standard ISO 9001. Internal audits are performed every year to train the use of its quality guidelines and to improve the quality level further on.

Research areas

By the end of 2015, IKET had employed more than 120 (full time equivalent) scientists, engineers, technicians and other personnel. Around half of the employees were funded in 2015 by the Helmholtz Association (HGF), the others by third party funds of the European Commission, by industry, by German ministries or by other research funds. Doctoral students as well as students of the Baden-Wuerttemberg Cooperative State University (DHBW) were filling around 20% of these positions at IKET. In addition, students perform their bachelor or master theses or spend an internship in the research laboratories of IKET.

IKET contributed in 2015 to the HGF programs "Nuclear Waste Management and Safety (NUSAFE)", "Renewable Energies (RE)", "Storage and Cross-Linked Infrastructures (SCI)", as well as "Nuclear Fusion (FUSION)". Fig. 2 shows the allocation of its personnel resources in 2015 to these programs. More than 2/3 of all IKET personnel was needed to solve issues on nuclear safety, on severe accident research or on alternatives for nuclear waste management.

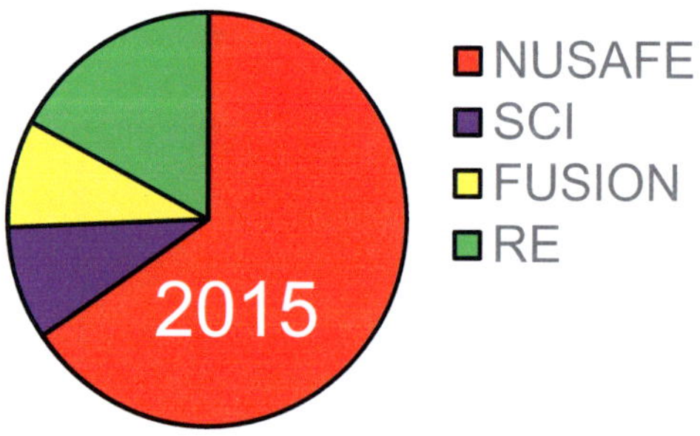

Fig. 2: Share of personnel resources of IKET in programs of the Helmholtz Association.

Technologies developed at IKET can be applied to a wide range of energy conversion systems. Liquid metal technologies, for example, can be used to cool advanced nuclear energy systems for transmutation of minor actinides in spent fuel, but they are also needed in blankets of nuclear fusion reactors, in targets of concentrated solar power systems, and even for cracking of methane to produce hydrogen as a carbon free gaseous fuel. The Helmholtz Alliance for Liquid Metal Technologies (LIMTECH) is supporting these synergies since 2015 with a joint research program, to which IKET is contributing with several doctoral theses.

The risk of hydrogen explosions with its disastrous consequences is not only to be mitigated during severe accidents of nuclear power plants, but needs to be managed as well for hydrogen fueled cars and their future infrastructures. Codes for flow and combustion of hydrogen, developed at KIT, are thus offered world-wide for containment design of nuclear power plants as well as for future mobile systems. A prominent example of the latter application is the hydrogen shuttle bus powered by fuel cells, which is operated by KIT and fueled at a gas station, which has been erected with technical support of IKET. Since 2015, IKET is contributing such hydrogen research to the new HGF program SCI.

The number of publications, as usual in research organizations, expresses the productivity of the institute. More than 200 publications, i.e. more than 3 per scientist and year, were given in 2015 to international journals, to conference proceedings and as book chapters.

IKET contributes to education and training along with scientific research with more than 1000 semester hours, not only on the university campus of KIT, but also at the Baden-Wuerttemberg Cooperative State University (DHBW), at the Hector School of KIT, in the AREVA Nuclear Professional School of KIT, and in Universities of Applied Sciences (FH). Moreover, IKET coordinates since 2015 the master courses on energy engineering at KIT.

The AREVA Nuclear Professional School is an IKET working group for education and training of young scientist in nuclear engineering, sponsored by AREVA GmbH since 2009.

Supported by lecturers of other organizations, this school has been offering compact courses in 2015 on nuclear technologies and methods. Moreover, this group is supervising doctoral students who were financially supported in 2015 by AREVA, RWE, VGB, and the European Commission. The contract with AREVA supporting the AREVA Nuclear Professional School has further been extended in 2015 until 2022.

Some Research Highlights of 2015

Funded by the Federal Ministry of Economic Affairs and Energy (BMWi), the Hydrogen Group has continued experimental work related to the flame behavior in partially confined layers of premixed gases with a concentration gradient. The focus in 2015 was on layers attached to vertical walls.

The experiments were conducted in the HYKA test vessel V30. The experimental matrix included homogeneous mixtures and pre-mixed layers with vertical and horizontal concentration gradients. Besides hydrogen inventory, ignition point and blockage ratio in the driver section have been varied, see Fig. 3.

In the SCI program, the Hydrogen Group focused on the bonfire test protocols and thermal shielding of pressure cylinders made from carbon fiber wrapped compound material and on the two-way coupling of the flame propagation and the structural dynamics of adjacent flexible walls.

Regarding the bonfire testing, it was shown that the quite vulnerable pressure cylinder material could be effectively protected applying an intumescent paint. This measure prolonged the time to catastrophic failure of the 700 bar loaded vessels by a huge factor. This would give sufficient time to first responders even in the unlikely event that the standard safety device, the temperature induced pressure relief device TPRD, did not activate.

Regarding the coupling of the flame to a flexible wall, a two way coupling of the KIT CFD code COM3d and the commercial FEM code ABAQUS was developed and tested. The Hydrogen Group conducted several related experiments in the HYKA Test Chamber Q160. The results showed good agreement with the respective numerical simulations.

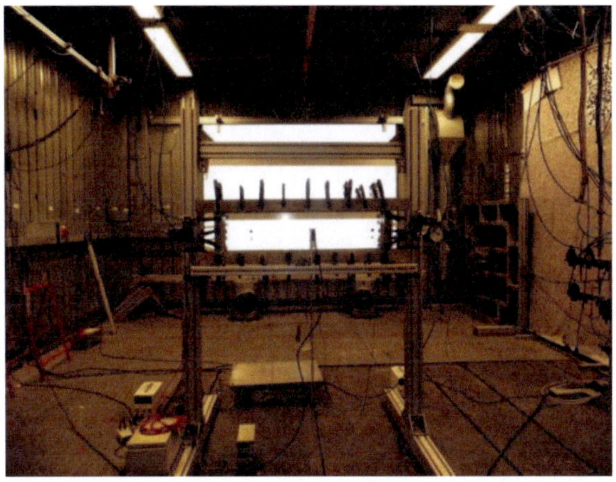

Fig. 3: Experimental set-up for investigating transient flame behavior coupled to flexible walls using transparent walls and optical measurement techniques in HYKA Test Chamber Q160.

The Energy and Process Engineering Group of IKET developed in 2015 a new measurement system to determine the thermo-physical properties of geothermal brines. Funded by BMWi, the national project PETher aims in determining the physical properties of brine under in-situ conditions, reflecting the original conditions in geothermal applications without precipitation of salts or release of gases. Fig, 4 shows the mobile device measuring the specific heat, the density and the kinematic viscosity of brine at various geothermal sites. A measurement system for thermal conductivity shall be added next.

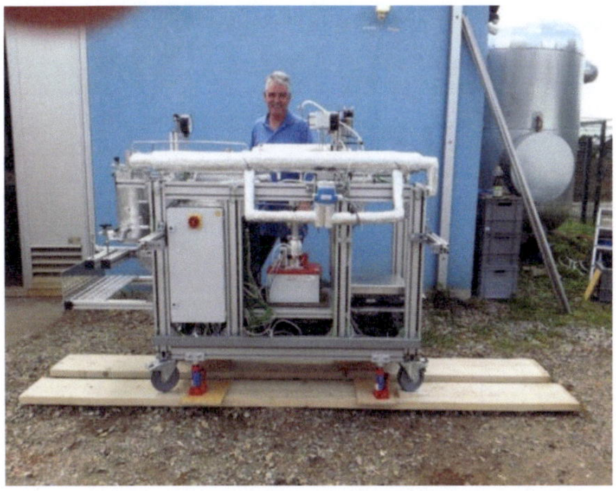

Fig. 4: In-situ measurement of specific heat, density and kinematic viscosity of geothermal brines, developed by the Energy and Process Engineering Group.

Production of hydrogen from methane without CO_2 emissions: What seems easy in theory - thermal cracking of methane into its constituents hydrogen and carbon - has not been achieved yet in a continuously running process. Reasons are the very high temperatures required, the tendency of carbon blocking any gas phase reactor by depositing at the walls or at any catalytically active surface, again raising the need for elevated temperatures. Following a proposal by Nobel Price laureate Prof. Carlo Rubbia and funded on his initiative by the Institute of Advanced Sustainability Studies IASS in Potsdam, the Karlsruhe Liquid Metal Laboratory investigated the option of using a liquid metal bubble column type reactor for cracking. After first successful trials in 2014, a major breakthrough has been achieved now by demonstrating a hydrogen yield of 78 %, while proving at the same time, that the carbon really is produced continuously as a powder in the process and separated in a relatively simple way by utilizing the density difference to the liquid metal. Fig. 5 shows the KALLA researchers together with Prof. Rubbia and his IASS colleagues during a joint project meeting at IKET.

Fig. 5: KALLA researchers together with Prof. Rubbia and his IASS colleagues during a joint project meeting on cracking of methane with liquid metals.

The study of liquid metal flows in strong magnetic fields plays an essential role in the development of nuclear fusion reactors, where breeding of tritium and heat extraction can be accomplished by circulating in the blanket the lithium-containing alloy PbLi. An existing scaled mock-up of a helium cooled lead lithium (HCLL) breeding blanket, which was available from previous analyses, has been upgraded such that the geometry near the plasma-facing first wall represents best the most recent design foreseen in ITER. Now the flow has to pass through multiple narrow gaps instead of a single larger opening that was present in an earlier design approach. Experiments have been performed in the MEKKA laboratory at IKET, Fig. 6, where the liquid metal flow has been investigated under the influence of a strong magnetic field. Measurements of magnetohydrodynamic pressure drop show that the new design creates about 3.5 times higher pressure drop near the first wall compared to experiments in the previous geometry. Nevertheless it turns out that even the increased pressure drop at the first wall remains moderate and acceptable in comparison with the one in the manifolds that distribute and collect the flow into the breeder units.

Fig. 6: Magneto-hydrodynamic flow test of a blanket module mock-up in the MEKKA laboratory.

The Accident Management Systems Group at IKET, together with the German aeronautics and space research centre (DLR) and the Research Center Jülich (FZJ), is part of the German Helmholtz research activity "Portfolio security". The three organisations combine their vast knowledge in cyber security, sensors and platforms, emergency management and security of critical infrastructures in one project. So far the central activity focuses on Critical Infrastructure Protection. An agent based modelling system was set-up to describe the interdependent critical infrastructures. As a starting point, agents were defined for the health care system (e.g. hospitals, elderly care and pharmacies) and water supply. Key parameters representing a critical infrastructure are the resources needed for its operation and the self-help capabilities in case of a lack of external resources. The city of Karlsruhe was selected as example location, as shown in Fig. 7. As next steps, households,

Fig. 7: Aerial representation of the affected area (light, partial blackout and dark, total blackout and loss of water supply); dots represent the status of a particular critical infrastructure.

emergency management organizations with their crisis teams and first responders will also be included. In addition, possible management options will be defined and implemented into the agent based system.

Passive safety system, which can refill cooling water into a boiling water reactor without the need of electric power, have become of particular interest after the Fukushima accident. As the reactor is still producing residual heat after shut-down, steam from the reactor can drive a turbine connected with a pump to refill water, as used in the Fukushima power plants, or it can simply be injected into cold water increasing its pressure beyond the reactor pressure. Such steam injectors had been used in the 19th century to refill water into the boiler of locomotives, but their application to nuclear power plants is still under discussion, as the physical flow phenomena inside the steam injector are rather complex. David Heinze, a young scientist of the AREVA Nuclear Professional School, completed his dissertation in 2015 on modelling the two-phase flow phenomena in such steam injectors using a one-dimensional theory. His predictions of atomization, of entrainment and mixing of steam and liquid, as well as of direct contact condensation of steam in cold water are describing surprisingly well the observed phenomena.

A large matrix of single-layer tests haven been performed in 2015 in the LIVE3D and LIVE2D facilities of the Accident Analysis Group to investigate the homogenous molten melt behavior in the lower head of a light water reactor under different severe accident scenarios and boundary cooling conditions. The test results were systematically analyzed in 2015 and the outcome of LIVE experiments provided more clarity for the remained uncertainties about the melt behavior. The melt flow regime, heat transfer efficiency and the heat flux distribution along the vessel wall have shown significant differences depending on the occurrence of crust formation, the vessel geometry in 3D or 2D and the combination of the top and external cooling conditions. Fig. 8 shows an infrared image showing the upward movement of hot cells on the melt surface.

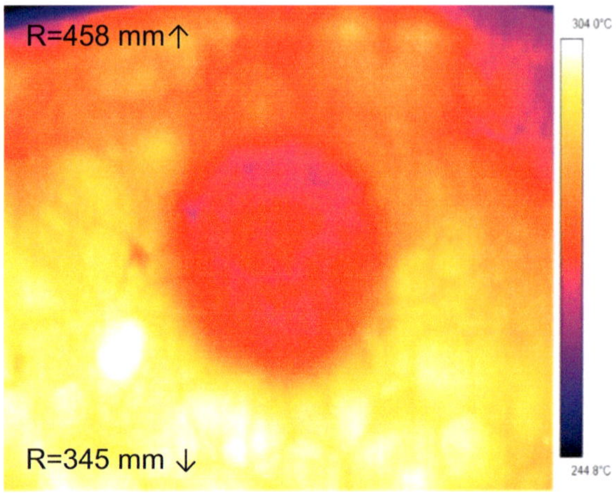

Fig. 8: Visualization of flow pattern in melt surface per infrared video images at LIVE3D facility

The Transmutation Group at IKET was concentrating in 2015 on severe accident simulations of nuclear systems with a fast neutron spectrum using the SIMMER code, which describes the interaction of fluid mechanics, neutron physics and structural damage of a degrading reactor core. For validation, KIT/IKET and Kyushu University (KU) participated in the IAEA project on simulation of transient experimental tests in the old sodium cooled fast reactor EBR-II. Results will further improve version III of the SIMMER code, which became the main tool for severe accident analyses in European projects on sustainable nuclear energy systems. All objectives of this KIT/KU cooperative work were achieved and several experiments were successfully simulated, as will be discussed in this report. The simulation results will also be reported at international conferences in 2016/2017 and will be included in an IAEA publication.

The following sections shall provide some deeper insight into selected research topics of these eight scientific working groups.

Magnetohydrodynamics for liquid-metal blankets

L. Bühler, T. Arlt, H.-J. Brinkmann, V. Chowdhury, C. Köhly, C. Mistrangelo

Influence of modifications of HCLL blanket design on MHD pressure losses

The study of liquid metal flows in strong magnetic fields plays an essential role in the development of nuclear fusion reactors where breeding of tritium and heat extraction can be accomplished by circulating in the blanket a lithium-containing alloy. The Helium Cooled Lead Lithium (HCLL) blanket is one of the concepts that will be tested in ITER. The eutectic alloy PbLi is used as breeder material and the heat is removed by helium flowing inside channels embedded in the walls. In the breeder zones the liquid metal flows at low velocity (<1mm/s) as needed to circulate the PbLi towards external ancillary systems for tritium removal and purification. A HCLL blanket module consists of a number of Breeder Units (BU) arranged in columns. Boxes are stiffened by a grid of plates to withstand the high helium pressure in case of an accidental in-box leak of coolant. This frame creates an array of rectangular cells where breeding units are positioned.

In 2008-2009 an experimental campaign was carried out to study MHD flows in a scaled mock-up of a HCLL blanket according to a design developed at CEA [1]. The test section consists of four BUs hydraulically connected two by two at the first wall (FW) through an opening that extended over the entire toroidal size of the BU (see Fig. 1(a)). A manifold feeds BU1 and BU3 and a second one collects the liquid metal from BU2 and BU4. The flow scheme is depicted in Fig. 2(a). Results showed that the major pressure drops occur in manifolds and at the first wall [2]. In order to improve the structural stability of the HCLL TBM the long gap at the first wall has been replaced by a series of small openings, as visualized in Fig. 1(b). The top view of one stiffening plate (SP) shows how the cross-section of the liquid metal flow path has been reduced due to the insertion of additional solid parts required to ensure mechanical stiffness. Therefore the velocity increases locally owing to the reduced cross-section along the flow path and the liquid metal flow contracts and expands along magnetic field lines to enter the upper BU.

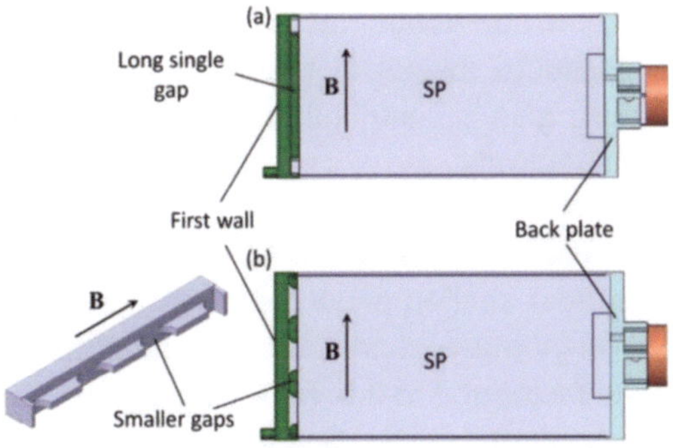

Fig. 1 (a): Original design of a stiffening plate (SP) between two BUs: at the FW the opening extends along the total width of the BU. (b) Modified design of the SP at the FW with a series of small openings.

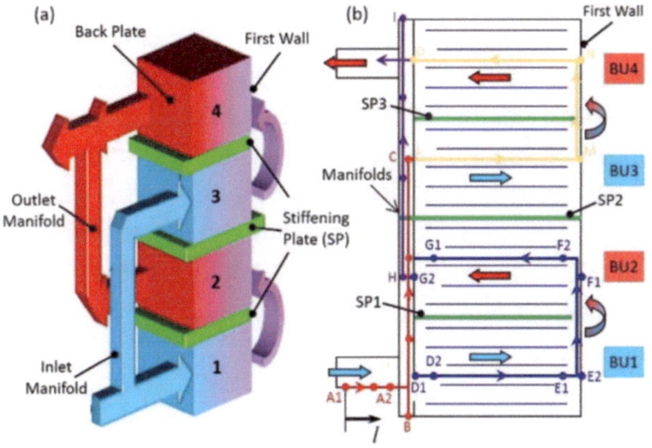

Fig. 2 (a): Scheme of liquid metal paths in the test section. The inlet manifold feeds BU1-BU3 and the second one collects the liquid metal from BU3-BU4. (b) Dots mark the position of pressure taps and arrows typical flow paths.

In order to integrate the described design modifications [1], the first wall of the former test section has been cut at locations corresponding to the stiffening plates that separate BU1 from BU2, and BU3 from BU4, and new parts have been inserted. From an MHD point of view the resulting flow distribution is expected to cause additional pressure drops due to the fact that the flow contracts and expands along magnetic field lines [3]. In order to quantify the influence of design modifications described above, the available test section has been changed to account for the new design features. Experiments have been performed to record pressure distribution in the new mock-up in a wide range of flow parameters and data have been compared with results obtained by using the former test section.

Pressure differences between pairs of pressure taps are measured via four capacitive pressure transducers mounted in series that are sensing the same pressure. Measurement spans of individual transducers overlap to avoid reading errors due to nonlinearity near the end of the ranges. From all four readings the one with the highest accuracy for the considered data is selected as *the* measured pressure value. For the discussion of the results the dimensional pressure p^* is normalized as $p = p^*/(\sigma u_0 L B^2)$. Here σ is the electric conductivity of the fluid, u_0 the average velocity in a cross-section of a breeder unit, L a typical length scale of the problem corresponding to half of the toroidal size of the mock-up, and B the magnitude of the external magnetic field. The pressure distribution is recorded along typical flow paths marked by different colors in Figure 2(b). Dots indicate pressure taps. The non-dimensional parameters characterizing the flow of a liquid metal with density ρ and kinematic viscosity ν are the Hartmann number Ha, the Reynolds number Re, and the interaction parameter N

$$Ha = BL\sqrt{\sigma/(\rho\nu)}, \quad Re = u_0 L/\nu, \quad N = Ha^2 / Re .$$

Pressure drops in geometric elements that form the test section are plotted in Fig. 3 as a function of N^{-1} for the flow at $Ha = 1000$. The main contributions to the total pressure drop (A1-I) occur in the inlet circular pipe Δp_{in} (A1-B), in the manifolds Δp_{M} (B-C, H-I), across the gap at the back plate Δp_{BP} (B-D1), and at the first wall Δp_{FW} (E2-F1). Δp_{BP} results from the expansion of the flow along magnetic field lines from the small manifold cross-section into the larger rectangular boxes. Δp_{FW} near the first wall is mainly caused by larger inertia forces for higher velocities, i.e. when $N^{-1} \gg 1$. The latter stem from the increased velocity due to the reduction of the cross-section along the flow path. Inside the breeder units pressure remains almost constant, since the liquid metal flows much slower than in pipes and manifolds. Results obtained with the modified test section (green, solid symbols) are compared with those recorded in the former experimental campaign carried out in 2009 (red and orange, open symbols in Fig. 3 und Fig. 4). For similar Reynolds numbers the pressure drop at the first wall is significantly larger when using the mock-up with small openings at the first wall.

For moderate Hartmann numbers all contributions depend on the flow rate. However, the pressure drop Δp_{FW} across the gap at the first wall exhibits the most significant dependence on the Reynolds number. It tends to zero for $Re/Ha^2 = N^{-1} \rightarrow 0$, suggesting that it is mainly related to the action of inertia forces. Instead, Δp_{BP} remains rather constant (8-12%) showing that it is determined primarily by electromagnetic phenomena. For $Ha = 1000$ in the parameter range investigated Δp_{FW} provides 5-35% of the total pressure drop for increasing Re. For larger magnetic field the pressure distribution in the mock-up is nearly unaffected by the flow rate, even for the highest Reynolds numbers reachable in the experiments (see Fig. 4). Even the losses at the FW are approximately constant and they provide about 1.5% of the total pressure drop. Pressure losses in manifolds and at the back plate give a contribution of 65-67% and 8-10%, respectively.

Experimental results obtained by using the modified test section show that design changes lead to an increase of pressure drop near the first wall by a factor of 3 to 3.5. As a consequence the total pressure drop becomes larger too. However, the influence of 3D MHD phenomena that occur at the FW are more relevant for moderate Hartmann numbers ($Ha < 2000$), since they are mainly related to inertia. For intense magnetic fields the contribution of the pressure drop at the FW wall is smaller than 2%. Therefore, even if these losses increase owing to the modified design, they are almost negligible compared to pressure drop in manifolds. It can be concluded that, although additional 3D MHD effects actually take place near the FW as a result of the expansion and contraction of the liquid metal when passing from one BU to the adjacent one, their influence on the total pressure drop is small especially for strong magnetic fields ($Ha > 2000$).

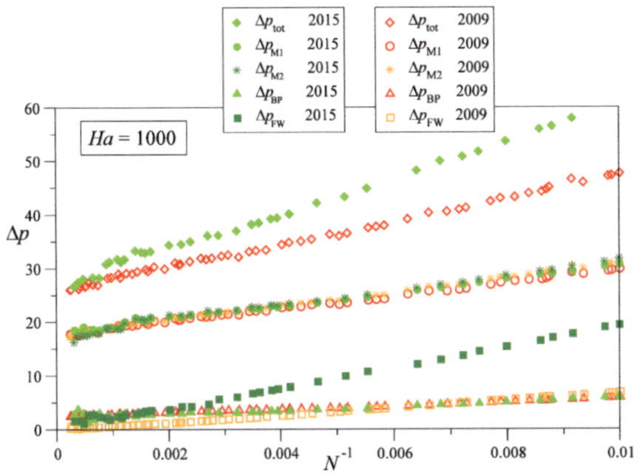

Fig. 3: Contributions to total pressure drop for Ha = 1000. Previous experimental data (2009) are compared with those obtained by using the modified test section (2015).

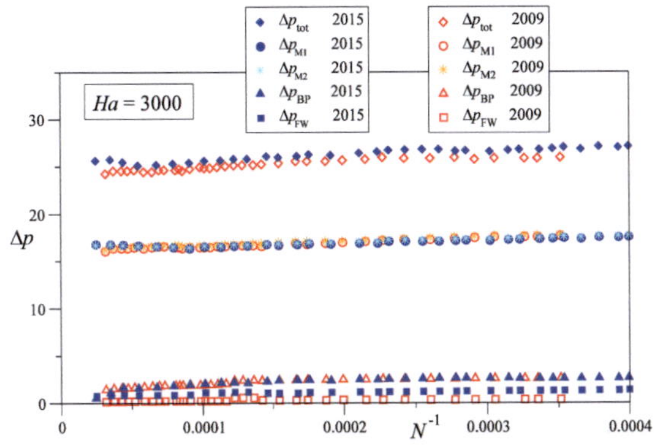

Fig. 4: Contributions to the total pressure drop for the flow at Ha = 3000. Results obtained with the modified test section are compared with the ones from the previous experimental campaign (2009).

Electromagnetic coupling of MHD flows.

In order to support design activities for a helium-cooled lead lithium (HCLL) blanket module for a DEMO reactor a numerical study is performed to simulate flow conditions in typical components of a HCLL blanket manifold (Fig. 5). Pressure distribution and occurrence of flow recirculation are investigated. Results depend on the position and conductivity of the wall that separates distributing and collecting manifolds and on the electromagnetic flow coupling between neighboring fluid regions. MHD flows are considered in a generic model geometry with typical features of the liquid metal manifold foreseen in a HCLL blanket module. The geometry, shown in Fig. 6, consists of two ducts which are electrically connected across a step-shaped dividing wall. As a result the flow undergoes periodic expansions and contractions and due to the electromagnetic coupling at the common separating wall, the flow in one duct influences the one in the neighboring channel. The half-length of one period in axial direction x is L and the alternating position of the dividing wall measured from the middle of the manifold is $\pm\Delta y$. When Δy=0 we have two electrically coupled straight square channels. In general the flow rates q_1 and q_2 in the two ducts may differ, depending on applied pressure differences.

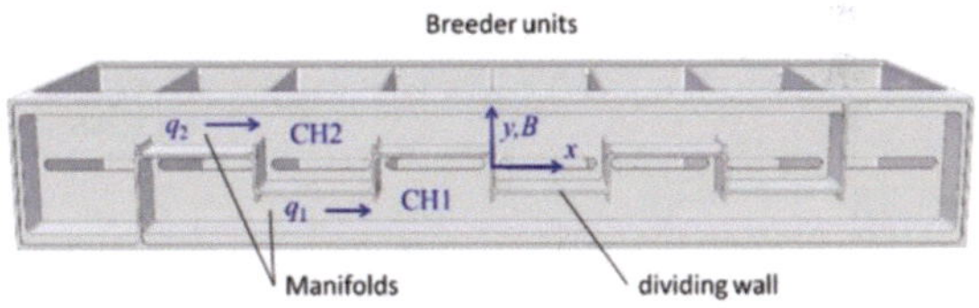

Fig. 5: MHD test section for studying the pressure drop in one PbLi manifold of a HCLL blanket module.

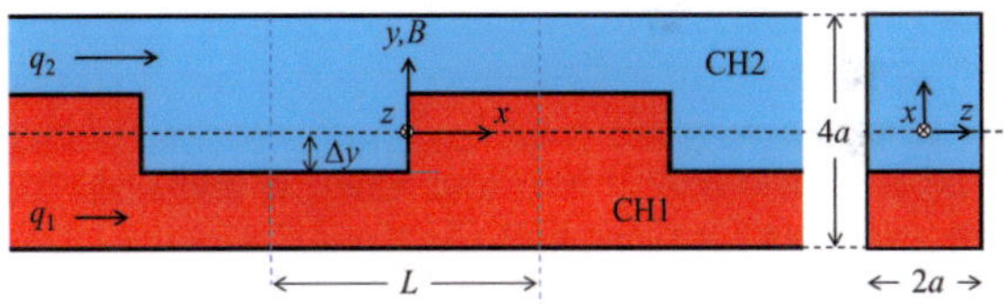

Fig. 6: Sketch of a generic manifold geometry. Two ducts with expansions and contractions are electrically coupled at a step-shape dividing wall.

Results are obtained by using a code based on an asymptotic method valid for large magnetic fields and small velocity (inertialess flow). For strong magnetic fields, viscous effects are confined to very thin boundary layers, while the core of the flow behaves practically as being inviscid. In addition complete numerical simulations for inertial MHD flows are performed using a code based on the open source package OpenFOAM. A more realistic geometry close to the one shown in Fig. 5 with walls of finite thickness and conductivity is considered.

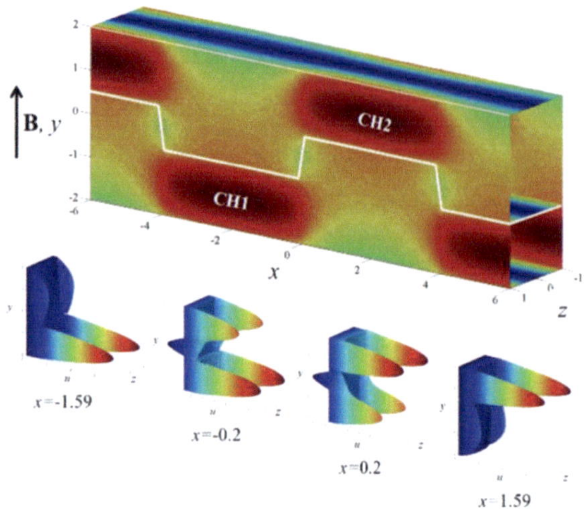

Fig. 7: Magnitude of surface potential plotted on the walls of the ducts for Ha=1000, $\Delta p_1=\Delta p_2$, and profiles of axial velocity at different axial positions.

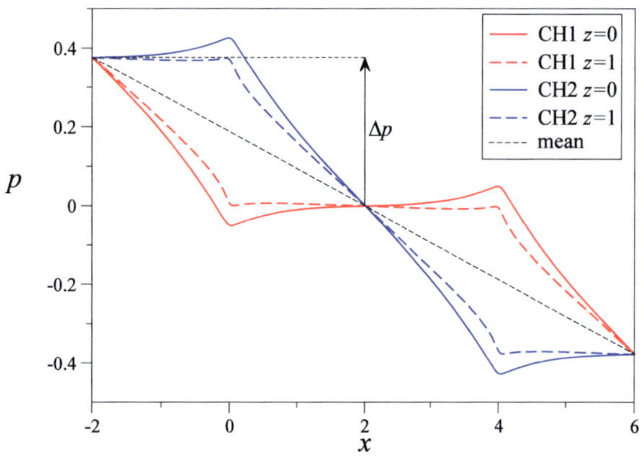

Fig. 8: Pressure along a full length of a period, plotted along a center line at z=0 and along a line near the side wall at z=1 for channels CH1 and CH2.

Results are shown for a reference case with expansion ratio Δy=0.5 and Ha=1000. Equal pressure differences are assumed in the two channels, $\Delta p_1=\Delta p_2$, and therefore both flow rates are equal, $q_1=q_2$. Surface electric potential is plotted in Fig. 7. In the narrow parts of the ducts the velocity is larger. This yields higher values of induced potential on the side walls at $z=\pm1$ (see dark red areas). Where the cross sections are larger the magnitude of induced potential difference is smaller. 3D MHD effects at expansions and contractions lead to modifications in the velocity distribution compared to fully developed conditions. This can be seen in subplots of Fig. 7, where profiles of axial velocity are displayed at several axial positions. At x=-1.59 the flow is almost fully developed with core velocities of order one and side layer jets along walls parallel to the magnetic field. At x=-0.2, before expansion of CH1 (contraction of CH2), a fraction of flow in the side layers moves in reversed direction. Behind the expansion of CH1 (contraction of CH2) at x=0.2 the highest velocities are observed in CH2 and a partly reserved flow occurs in the layers of CH1. At x=1.59 the flow is again almost fully developed. In the narrow parts of the ducts induced currents are large and as a result we find here the strongest pressure gradients. This can be seen in Fig. 8, e.g. for CH1 in the regions $-2<x<0$ and $4<x<6$ or for CH2 in the range $0<x<4$. Near expansions and contractions the magnitude of pressure gradients increases due to additional Lorentz forces caused by 3D electric currents. In the small channels, where velocities are large, induced currents and braking Lorentz forces are high. Here mechanical energy is removed from the flow and transformed into electrical energy. The latter is transferred via leakage currents through conducting walls to the neighboring larger ducts and released there partly as mechanical energy to drive the slower flow in the larger cross section. The small ducts act as local electric MHD generators while the larger ducts function as electromagnetic pumps. This can be seen by the increasing pressure with respect to the mean value when moving along the streamwise direction in CH1 in the range $0<x<4$ and for CH2 in the regions $-2<x<0$ and $4<x<6$. The pressure drop Δp over a half-length of a period, as indicated in Fig. 8, has been investigated for different expansion ratios Δy. For Δy=0 we observe the pressure drop of a fully developed flow in two straight, electrically coupled channels. With increasing Δy the pressure drop increases for two reasons. First, the flow has to pass through smaller cross sections. This leads to increased velocity and higher pressure drop. Second, larger expansion ratios increase the intensity of 3D MHD effects associated with additional pressure drop. For Δy=0.6, which is still a bit smaller than that in the blanket, the increase in Δp is already 91% compared to a flow in two parallel channels. First results have been published in [4].

Analytical study of instabilities in magnetohydrodynamic boundary layers

Experiments to investigate instabilities in fundamental MHD flows are performed, both to obtain data for validation of computational codes and to reach a deep understanding of basic MHD phenomena needed for the physical interpretation of complex, coupled problems. Experiments have been carried out in the MEKKA laboratory to analyze liquid metal flows in rectangular channels under the influence of intense magnetic fields. The investigated MHD flows are characterized by the two non-dimensional groups, the Hartmann number Ha, describing the magnetic field strength, and the Reynolds number Re, describing the average axial velocity, and by the wall conductance ratio $c=\sigma_w t_w/(\sigma L)$ for walls with electric conductivity σ_w and thickness t_w. The conductance parameter c describes the ratio of the conductance of wall material to the one of the fluid. In the test section used for the experimental campaigns walls parallel to the magnetic field are electrically insulating ($c=0$) and those perpendicular are covered by thin foils of copper to achieve a well-defined electric wall conductance $c>0$. These electric boundary conditions correspond to the classical Hunt-type MHD flow [4]. For strong magnetic fields (Ha>>1) stationary laminar MHD flows in electrically conducting ducts exhibit a uniform core velocity with thin viscous Hartmann layers at walls perpendicular to the magnetic field and high-velocity jets along walls aligned with the field. These jet-like velocity patterns may become unstable for high enough flow rates, i.e. for large Reynolds numbers [5].

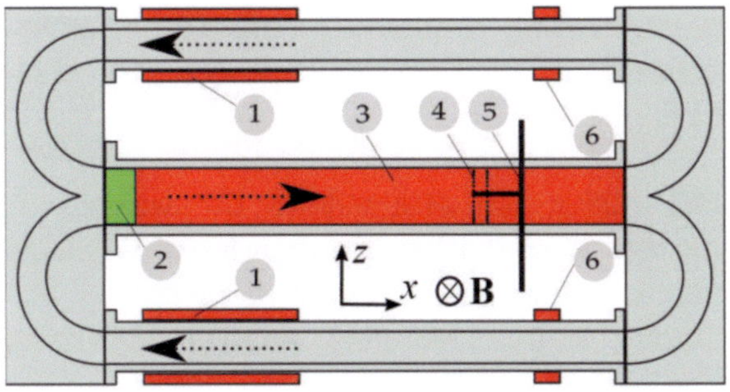

Fig. 9: GaInSn loop with (1) electromagnetic pumps, (2) flow straightener, (3) test section with conducting Hartmann walls, (4) potential probes on duct wall, (5) traversable probe and (6) flow meters.

The design of the employed liquid metal loop is shown in Fig. 9. Two lateral ducts serve as electromagnetic conduction pumps that feed the test section in the middle. The eutectic alloy GaInSn is used as model fluid and the entire loop is exposed to a uniform magnetic field.

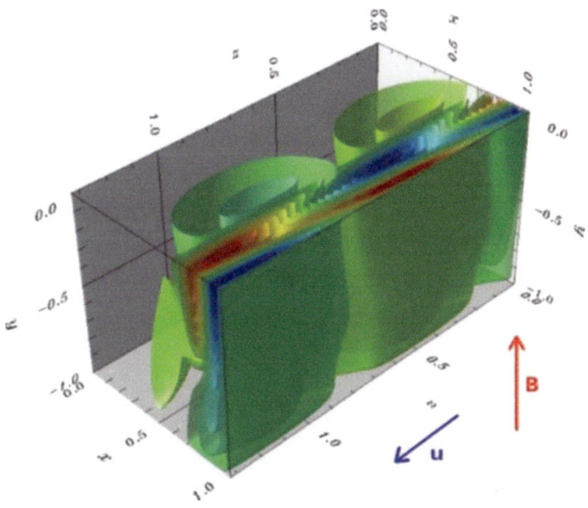

Fig. 10:. Perturbation of axial velocity component for one wavelength in the lower right quarter for c=1, Rec=260 and Ha=100.

In addition to the experimental studies a linear stability analysis has been performed for liquid-metal flows in square ducts with thin electrically conducting walls in a homogeneous magnetic field. A vector streamfunction-vorticity formulation is used to describe the problem in terms of perturbation equations. The resulting eigenvalue problem for three-dimensional small-amplitude perturbations is solved by a Chebyshev collocation method. Compared to ideal assumptions used in previous studies [5] the analysis has been extended to deal with realistic wall conductance ratios $c=1$, 0.1, 0.01. Calculations have been performed for Hartmann numbers up to 10^4 [6]. As expected for MHD duct flows in sufficiently strong magnetic fields instabilities occur first in the side layers, whose thickness is proportional to $Ha^{-1/2}$, when the Reynolds number exceeds a critical value Re_c. The Reynolds number is defined in this analysis by the maximum velocity in the jets. Fig. 10 shows perturbations of axial velocity component in the lower right quarter of the duct for one wavelength for $c=1$, $Re_c=260$ and $Ha=100$. Perturbations are located in the side layers and they are aligned along magnetic field lines. At high Hartmann numbers, an asymptotic behavior for the critical Reynolds number is observed, i.e. $Re_c \approx 520$ for $Ha >> 1$. The axial wave length of unstable struc-tures scales proportionally to the thickness of the jets since $k_c \sim 0.5 Ha^{1/2}$. The present results for Re_c are still lower compared to experimental data but larger than the value of $Re_c = 313$ found in [7].

References

[1] Rampal, G., Li-Puma, A., Poitevin, Y., Rigal, E., Szczepanski J. and Boudot, C., "HCLL TBM for ITER - design studies," *Fusion Engineering and Design,* Vols. 75-79, pp. 917-922, 2005.

[2] Mistrangelo C. and Bühler, L., "Magneto-hydrodynamic pressure drops in geometric elements forming a HCLL blanket mock-up," *Fusion Engineering and Design,* vol. 86, pp. 2304-2307, 2011.

[3] Bühler, L., "Three-dimensional liquid metal flows in strong magnetic fields," Technical Report FZKA 7412, Forschungszentrum Karlsruhe, 2008.

[4] Bühler L. and Mistrangelo, C., "Theoretical Studies of MHD Flows in Support to HCLL Design," *Fusion Engineering and Design,* p. accepted, 2016.

[5] Hunt, J. C. R., "Magnetohydrodynamic flow in rectangular ducts," *Journal of Fluid Mechanics,* vol. 21, pp. 577-590, 1965.

[6] Priede, J., Aleksandrova, S. and Molokov, S., "Linear stability of Hunt's flow," *Journal of Fluid Mechanics,* vol. 649, pp. 115-134, 2010.

[7] Priede, J., Arlt T. and Bühler, L., "Linear stability of magnetohydrodynamic flow in a square duct with thin conducting walls," *Journal of Fluid Mechanics,* vol. 788, pp. 120-146, 2016.

[8] Ting, A. L., Walker, J. S., Moon, T. J., Reed C. B. and Picologlou, B. F., "Linear stability analysis for high-velocity boundary layers in liquid-metal magnetohydrodynamic flows," *Int. J. Engng. Sci.,* vol. 29, no. 8, pp. 939-948, 1991.

Read more:

[9] Aiello, G.; Mistrangelo, C.; Bühler, L.; Mas de les Valls, E.; Aubert, J.; Li-Puma, A.; Rapisarda, D.; Del Nevo, A. (2015). *MHD issues related to the use of Lithium Lead eutectic as breeder material for blankets of Fusion power plants.* *Magnetohydrodynamics*, 51, 185-194.

[10] Arlt, T.; Bühler, L.; Priede, J. (2015) *Numerical investigation of magnetically induced instabilities.* 16th MHD Days, Ilmenau, December 7-9, 2015.

[11] Bühler, L.; Chowdhury, V.; Mistrangelo, C.; Brinkmann, H. J. (2015). *MHD instabilities in boundary layers aligned with a magnetic field.* 6th International Symposium on Bifurcations and Instabilities in Fluid Dynamics (BIFD 2015), Paris, F, July 15-17, 2015 Book of Abstracts.

[12] Bühler, L.; Mistrangelo, C. (2015). *Three-dimensional MHD effects due to discontinuous insulation in pipes and channels.* 16th MHD Days, Ilmenau, December 7-9, 2015.

[13] Bühler, L.; Mistrangelo, C. (2015). *Theoretical studies of MHD flows in support to HCLL design activities.* 12th International Symposium on Fusion Nuclear Technology (ISFNT 2015), Jeju, Korea, September 14-18, 2015.

[14] Bühler, L.; Mistrangelo, C.; Konys, J.; Bhattacharyay, R.; Huang, Q.; Obukhov, D.; Smolentsev, S.; Utili, M. (2015). Facilities, testing program and modeling needs for studying liquid metal magnetohydrodynamic flows in fusion blankets. *Fusion Engineering and Design*, 100, 55-64. doi:10.1016/j.fusengdes.2014.03.078

[15] Bühler, L.; Mistrangelo, C.; Molokov, S. (2015). Validity of quasi-2D models for magneto-convection. *Magnetohydrodynamics*, 51, 321-327.

[16] Bühler, L.; Mistrangelo, C.; Najuch, T. (2015). Magnetohydrodynamic flows in model porous structures. *Fusion Engineering and Design*, 98-99, 1239-1243. doi:10.1016/j.fuseng des.2015.01.018

[17] Chowdhury, V.; Bühler, L. (2015). *Experimental investigation of magnetically induced instabilities (A3).* LIMTECH Summer School, Karlsruhe, June 30 - July 1, 2015.

[18] Chowdhury, V.; Bühler, L. (2015). *Calibration of potential probes for measuring velocity profiles in liquid metal MHD duct flows*. 3rd International Workshop on Measuring Techniques for Liquid Metal Flows (MTLM 2015), Dresden, April 15-17, 2015.

[19] Chowdhury, V.; Bühler, L.; Mistrangelo, C.; Brinkmann, H. J. (2015). Experimental study of instabilities in magnetohydrodynamic boundary layers. *Fusion Engineering and Design*, 98-99, 1751-1754. doi:10.1016/j.fusengdes.2015.02.027

[20] Chowdhury, V.; Bühler, L.; Mistrangelo, C.; Brinkmann, H. J. (2015). *Experimental investigation of unstable magnetohydrodynamic boundary layers*. EST 2015 - Energy, Science and Technology, Karlsruhe, May 20-22, 2015.

[21] Kessel, C.E.; Tillack, M.S.; Najmabadi, F.; Poll, F.M.; Ghantous, K.; Gorelenkov, N.; Wang, X.R.; Navael, D.; Toudeshki, H.H.; Koehly, C.; El-Guebaly, L.; Blanchard, J.P.; Martin, C.J.; Mynsburge, L.; Humrickhouse, P.; Rensink, M. E.; Rognlien, T.D.; Yoda, M.; Abdel-Khalik, S. I.; Hageman, M.D.; Mills, B.H.; Rader, J. D.; Sadowski, D.L.; Snyder, B.; John, H.S.; Turnbull, A.D.; Waganer, L.M.; Malang, S.; Rowcliffe, A. F. (2015). The ARIES advanced and conservative tokamak power plant study. *Fusion Science and Technology*, 67, 1-21. doi:10.13182/FST14-794

[22] Krasnov, D.; Braiden, L.; Molokov, S.; Boeck, T.; Bühler, L. (2015). *MHD flow instability in ducts with conducting walls*. 6th International Symposium on Bifurcations and Instabilities in Fluid Dynamics (BIFD 2015), Paris, F, July 15-17, 2015 Book of Abstracts.

[23] Mistrangelo, C.; Bühler, L. (2015). *MHD velocity jets driven by injected currents*. 16th MHD Days, Ilmenau, December 7-9, 2015.

[24] Mistrangelo, C.; Bühler, L. (2015). Magneto convective instabilities driven by internal uniform volumetric heating. *Magnetohydrodynamics*, 51, 303-310.

[25] Mistrangelo, C.; Bühler, L. (2015). *MHD instabilities in electrically driven rotating jets*. 6th International Symposium on Bifurcations and Instabilities in Fluid Dynamics (BIFD 2015), Paris, F, July 15-17, 2015 Book of Abstracts.

[26] Mistrangelo, C.; Bühler, L. (2015). *Electromagnetic flow coupling for liquid metal blanket applications*. 12th International Symposium on Fusion Nuclear Technology (ISFNT 2015), Jeju, Korea, September 14-18, 2015.

[27] Reimann, J.; Brun, E.; Ferrero, C.; Vicente, J. (2015). Pebble bed structures in the vicinity of concave and convex walls. *Fusion Engineering and Design*, 98-99, 1855-1858. doi:10.1016 /j.fusengdes.2015.05.026

[28] Reimann, J.; Ferrero, C.; Vicente, J.; Rack, A.; Brun, E.; Gan, Y. (2015). *Characterization of structured and non-structured zones of pebble beds, results from recent microtomography experiments*. Vortr.: University of California, Los Angeles, Calif., 7. Oktober 2015.

[29] Reimann, J.; Fretz, B.; Pupeschi, S. (2015). Thermo-mechanical screening tests to qualify beryllium pebble beds with non-spherical pebbles. *Fusion Engineering and Design*, 98-99, 1851-1854. doi:10.1016/j.fusengdes.2015.04.046

[30] Schulenberg, T. (2015), *Annual Report 2014 of the Institute for Nuclear and Energy Technologies (KIT Scientific Reports; 7702)*. KIT Scientific Publishing, Karlsruhe. doi:*10.5445/KSP/1000047907*

[31] Smolentsev, S.; Badia, S.; Bhattacharyay, R.; Bühler, L.; Chen, L.; Huang, Q.; Jin, H.G.; Krasnov, D.; Lee, D.W.; Mas de les Valls, E.; Mistrangelo, C.; Munipalli, R.; Ni, M.J.; Pashkevich, D.; Patel, A.; Pulugundla, G.; Saty-amurthy, P.; Snegirev, A.; Sviridov, V.; Swain, P.; Zhou, T.; Zikanov, O. (2015). An approach to verification and validation of MHD codes for fusion applications. *Fusion Engineering and Design*, 100, 65-72. doi:10.1016/j.fusengdes.2014.04.049

[32] Wang, X.R.; Tillack, M.S.; Toudeshki, H.H.; Koehly, C.; Malang, S.; Najmabadi, F.; ARIES Team. ARIES-ACT2 DCLL power core design and engineering. *Fusion Science and Technology*, 67, 193-219. doi:10.13182/FST14-798

[33] Wang, X. R.; Tillack, M. S.; Koehly, C.; Malang, S.; Toudeshki, H. H.; Najmabadi, F.; ARIES Team. (2015). ARIES-ACT1 system configuration, assembly, and maintenance. *Fusion Science and Technology*, 67, 22-48. doi:10. 13182/FST14-797

Group: Transmutation

Benchmark Analyses of EBR-II Shutdown Heat Removal Tests

Barbara Vezzoni, Marco Marchetti, Fabrizio Gabrielli, Lena Andriolo, Xue-Nong Chen, Vladimir Kriventsev, Claudia Matzerath Boccaccini, Andrei Rineiski, Werner Maschek

Introduction

In June 2012, the International Atomic Energy Agency (IAEA) initiated a four-year Coordinated Research Project (CRP) with the objective of improving state-of-the-art liquid metal cooled fast reactor codes and data used in neutronics, thermal hydraulics and safety analyses, by considering validation against whole-plant data recorded during landmark shutdown heat removal tests (SHRT) that were conducted at Argonne's Experimental Breeder Reactor II (EBR-II) in the 1980's. Twenty-two organizations representing eleven countries are participating in the CRP under the technical leadership of the Argonne National Laboratory (ANL). As secondary CRP aim, the training of the next generation of fast reactor analysts through international benchmark exercises is considered.

For the study, two of the tests performed at the EBR-II plant – a protected loss-of-flow (PLOF, SHRT-17) and an unprotected loss of flow (station blackout, ULOF, SHRT-45R), both initiated from full power and flow, were considered. During the analysis of the protected loss-of-flow, the focus was mainly on prediction of the natural convective cooling, while during the analysis of the unprotected loss-of-flow, more attention was devoted to the feedback effects as the thermal expansion reactivity effects. In order to support these studies, an optional neutronics benchmark analysis was considered as part of the SHRT-45R simulations.

The CRP activity was divided into different phases. During phase 1, the neutronics benchmark was carried out and the participants generated blind simulation results for both transients. After the distribution of the recorded experimental data, phase 2 started by performing several activities for improving simulation results. The participants decided also to apply the Best Estimate plus Uncertainties (BEPU) methodology (up to now used only in Light Water Reactors – LWRs - framework) for the qualification of the EBR-II simulation (PLOF test).

The KIT/IKET (TRANS group) is contributing to all the phases by performing transient analyses with the SIMMER code and neutronics investigations by means of the codes ERANOS 2.2 and PARTISN v5.97.

EBR-II reactor and core layout

The EBR-II plant was a uranium metal-alloy-fuelled sodium-cooled fast reactor (62.5 MWth) designed and operated by ANL for the U.S. Department of Energy at Argonne-West. Operation began in 1964 and continued until 1994. The original emphasis in the design and operation of EBR-II was to demonstrate the feasibility of a closed fuel cycle (breeding Pu) with on-site pyro-metallurgical reprocessing. Later on, it has been dedicated to safety studies by performing several Shutdown Heat Removal Tests (SHRTs) and Inherent Safety Tests (ISTs).

As indicated in Figure 1, EBR-II was a pool type reactor in which all major primary system components were submerged. Two primary pumps drew sodium from this pool and provided sodium to the two inlet plena for the core by a piping system. Subassemblies (SAs) in the inner core and extended core regions received sodium from the high-pressure inlet plenum accounting for approximately 85% of the total primary flow. The blanket and reflector subassemblies in the outer blanket region received sodium from the low-pressure inlet plenum. Hot sodium exited the SAs into a common upper plenum where it is mixed before passing through the outlet pipe (so-called "Z-pipe") into the intermediate heat exchanger (IHX). Sodium then exited the IHX flowing back into the primary sodium tank before entering the primary sodium pumps again.

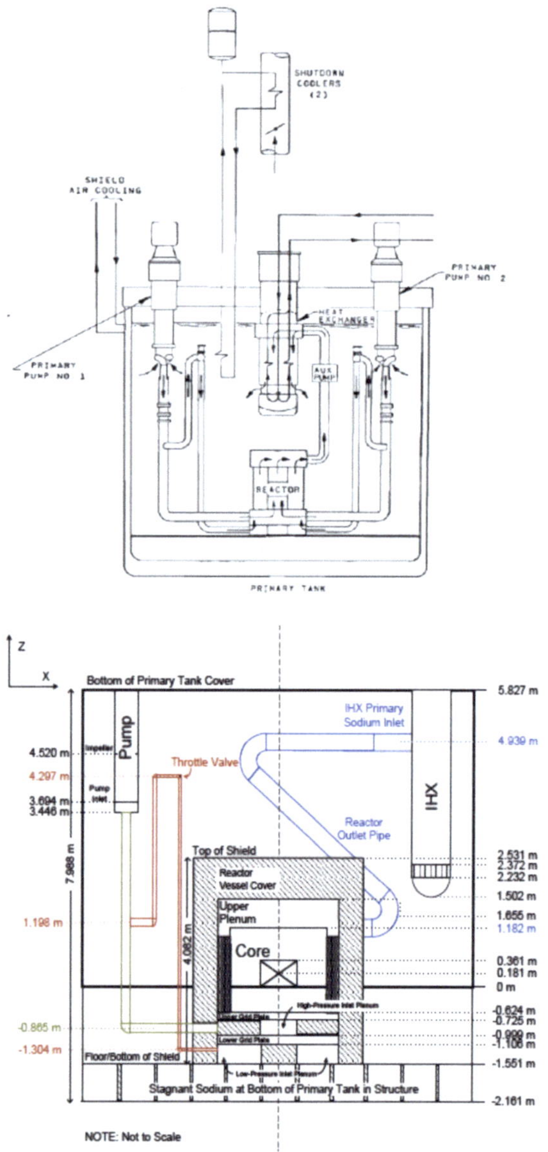

Fig. 1: EBR-II Primary System Components and Sodium Flow Paths [10]

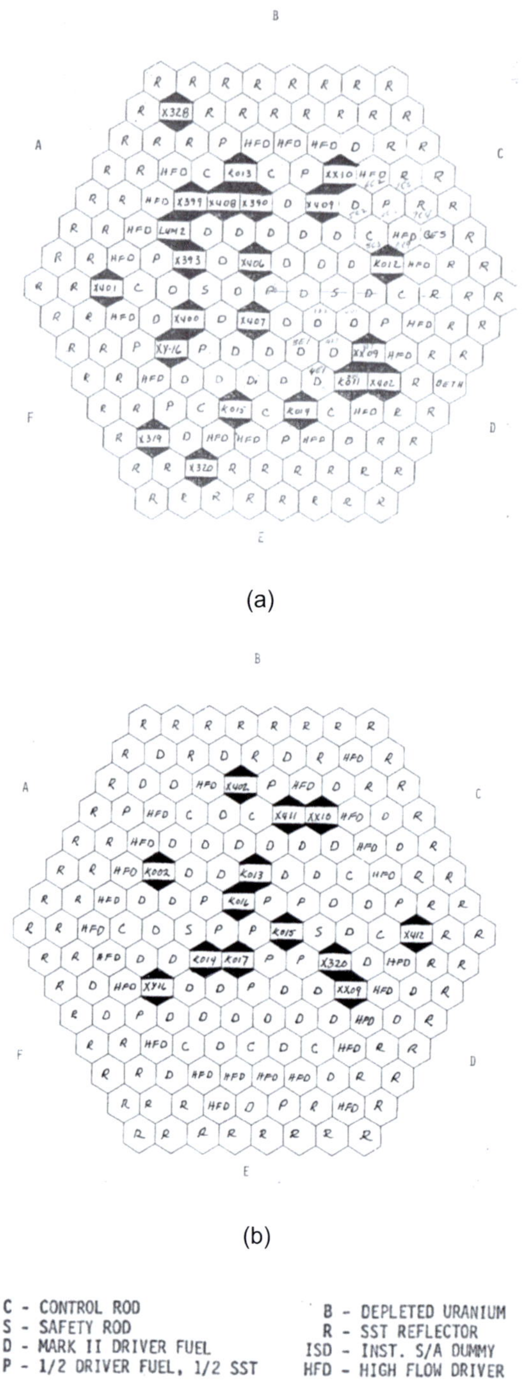

Fig. 2: Core layout: (a) SHRT-17 – PLOF (b) SHRT-45R – ULOF [10]

The core layout for the two considered tests was different as indicated in Figure 2. Therefore, in order to take into account the peculiarities of the two tests (several SA types composing the core), two different SIMMER core models were established. For the PLOF case, the core zone has been modeled by 19 radial meshes while for the ULOF case 34 radial meshes have been considered.

All the reactor components have been taken into account for creating the 2D RZ SIMMER-III model (Figure 3). Unavoidable approximations (necessary for 2D geometry) have been introduced for modeling the reactor outlet Z-pipe and the sodium inlet pipes. In addition, a simplified Intermediate Heat Exchanger (IHX) model (characterized by a constant sodium outlet temperature) and a single “equivalent” pump (representing the two primary pumps and the Electromagnetic Pump) have been implemented in the model.

In order to carry out the EBR-II benchmark studies, the standard SIMMER-III version has been modified by Kyushu University (KU) for taking into account a set of specific Equations of State (EOS) and the Thermo-Physical Properties (TPP) for the EBR-II fuel (U-5%Fissium alloy).

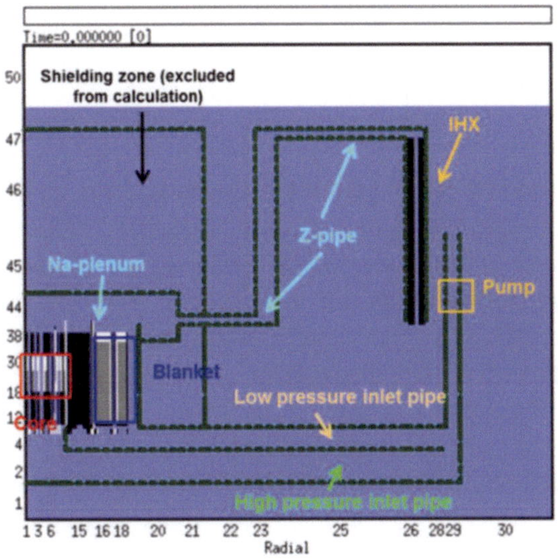

Fig. 3: SIMMER-III EBR-II RZ model (PLOF SHRT-17)

EBR-II results: comparison against experimental data

The PLOF case

For SHRT-17 – PLOF, oriented mainly to investigate the effectiveness of natural circulation, only the fluid-dynamics modules of SIMMER were considered (neutronics module deactivated) resulting in a reduced calculation time. Under these favorable calculation time conditions, several modeling options (e.g. radial conduction, axial conduction, effect of IHX position, gap conductions, fuel porosity, etc.) were investigated. The results obtained during phase 2 (optimization phase) are in good agreement with the experimental data as indicated in Figure 4 where the Z-pipe inlet temperature and the coolant temperature at core top in the instrumented SA XX10 (steel SA) are shown as examples. The improvement for XX10 is mainly related to the radial heat condiction.

The ULOF case

In order to establish a neutronics reference for SIMMER simulations, KIT/IKET extensively contributed to the neutronics benchmark part before performing ULOF transient simulations. Several codes (ERANOS 2.2, PARTISN v5.97, MCNPX 2.7.0) and modeling options (heterogeneous/homogeneous models for generation of multi-group cross-sections, diffusion/transport approximations for neutron transport, etc.) were considered and compared. A very good agreement with the results obtained by the other participants has been obtained. An example of parametric studies performed for the neutronics benchmark is shown in Table 1. More details will be published in [5].

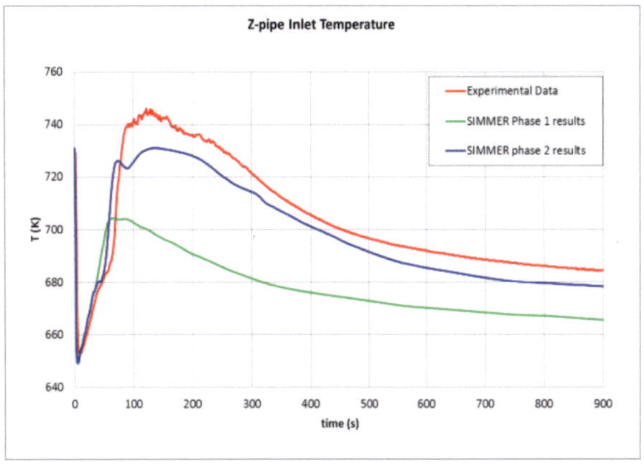

(a)

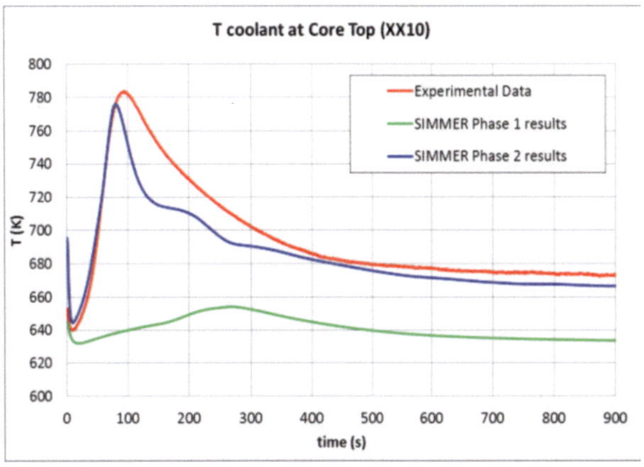

(b)

Fig. 4: SHRT-17 – PLOF: (a) Z-pipe Inlet Temperature (b) coolant temperature at core top (XX10)

Table 1: EBR-II SHRT-45R: results neutronics benchmark

	ERANOS						PARTISN
	HEX-Z Cold dimensions (REF)	**HEX-Z Expanded dimensions**	**HEX-Z Homogeneous XS treatment**	**HEX-Z P3**	**HEX-Z Diffusion**	**XYZ**	**Sn-16**
k-eff	0.99969	0.98761	0.99751	1.00198	0.97482	0.99960	1.00339
Time (s)	409	-	-	3305	-	6380	62
	Effects on k-eff of modeling options						
Effect, pcm	-	-1224	-219	229	-2552	-9	369

In addition to the specific EOS and TPP introduced in SIMMER by KU, extended SIMMER versions have been developed at KIT/IKET and adopted in the ULOF study. The most important modifications introduced are: 1) a new core thermal expansion reactivity feedback model developed at KIT [8-10], and 2) a new PARTISN-based spatial kinetics model (instead of a TWODANT-based one) used in order to benefit from the SIMMER parallelization capability. The comparison against whole-plant data allows a further validation of the SIMMER extensions considered.

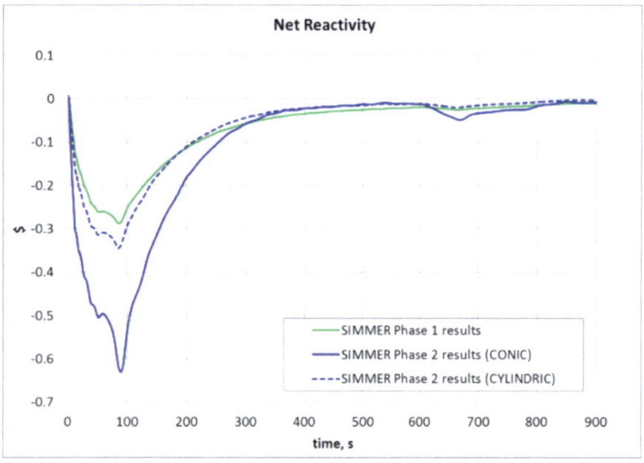

Fig. 5: SHRT-45R – ULOF: net reactivity [6]

In the new core thermal expansion reactivity feedback model introduced in SIMMER, axial core expansion considers fuel or clad driven modes (depending on the burnup level, namely depending on the state of the gap closure) as well as radial core expansion considers both cylindrical (only grid plate expands) and conic (both grid and constraints plates expand) modes. Within the study, both options have been considered as shown in Figure 5. The results obtained by the conic mode are in good agreement with the results presented by other participants.

Comparison with several available experimental data (including those in the XX09 and XX10 instrumented SAs) has been performed as well. As an example, Figure 6 shows the coolant temperature behavior at different axial positions in the fueled instrumented SA (XX09). The agreement with the experimental data is quite good for the temperature at core top (at 32.2 cm from core bottom, Figure 6-a) and for the temperature above core (at 48 cm from the core bottom, Figure 6-b). Discrepancies at the long term (after 600 s) are due to the missing contribution to the mass flow coming from the Electromagnetic Pump (EM) installed outside of the Z-pipe which is not properly modeled in our current study. This contribution has not been initially considered due to the simplified pump model adopted in SIMMER (a single "equivalent" pump) but it is currently under investigation in view of the submission of the final results.

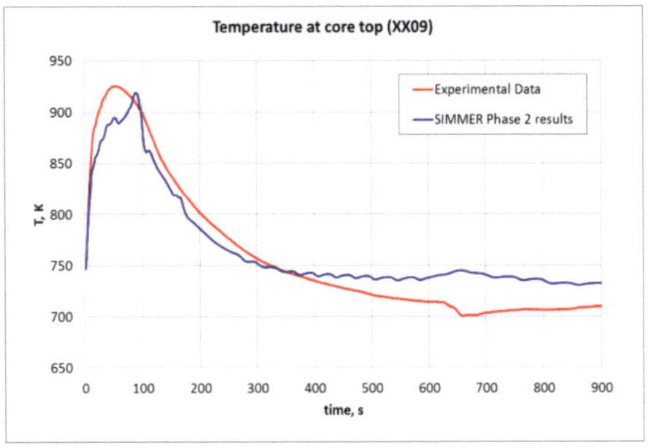

(a)

Fig. 6 a): SHRT-45R – ULOF (conic mode): XX09 temperature at core top

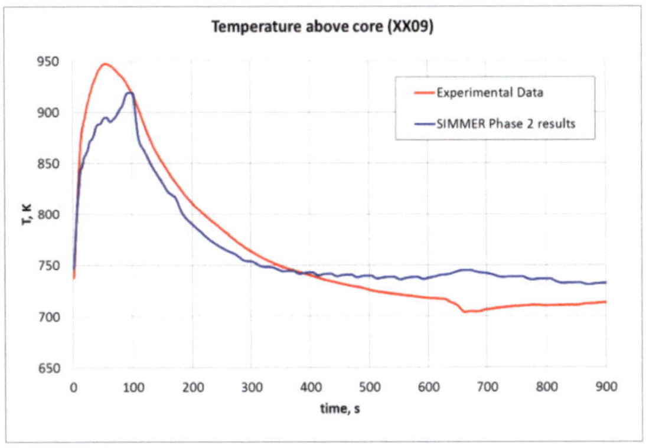

(b)

Fig. 6 b):. SHRT-45R – ULOF (conic mode): XX09 temperature above core

Conclusions

The group TRANS of KIT/IKET is participating in the IAEA EBR-II benchmark activity since June 2012, by performing transient analyses with the SIMMER-III code and neutronics investigation with the ERANOS and PARTISN codes.

Analyses for two tests (SHRT-17 - PLOF and SHRT-45R - ULOF) have been performed after having assessed two different RZ models for taking into account the core configurations of the two tests. An extended SIMMER version has been developed at KIT/IKET and validated during the study. In this extended version, specific EOS and TPP for the EBR-II fuel have been implemented by KU and a new core thermal expansion reactivity feedback model has been developed by KIT/IKET [9]. The version adopted also includes a new PARTISN-based spatial kinetics model in order to benefit from the parallelization capabilities of PARTISN and SIMMER codes. Contribution to the neutronics benchmark part has also been provided in order to establish an independent neutronics reference for SIMMER simulations.

The results obtained in phase 1 and 2 for the two tests are within the range of results obtained by other benchmark participants. By model refinements (phase 2), oriented to a better understanding of the phenomena involved, it has been possible to obtain a good agreement with the experimental data both for the PLOF and ULOF cases. Therefore, the study has provided a basis for a further validation of the SIMMER code.

References

[1] Briggs, L., Choi, C., Hu, W., Maas, L., Maschek, W., Merk, B., Mikityuk, K., Mochizuki, H., Monti, S., Morita, K., del Nevo, A., Ohira, H., Petruzzi, A., Sarathy, U., Shin, A., Shvetsov, I., Stempniewicz, M., Su, G., Sui, D., Truong, B., "Benchmark Analyses of the Shutdown Heat Removal Tests Performed in the EBR-II Reactor", Proc. Int. Conf. FR13, Paris, France, March 4-7, 2013.

[2] Fei, T., Vezzoni, B., Marchetti, M., van Rooijen, W.F.G., Zhang, Y., "Neutronics Benchmark for EBR-II Shutdown Heat Removal Test SHRT-45R", to be presented at the Int. Conf. PHYSOR 2016, Sun Valley, Idaho, USA, May 1 – 5, 2016.

[3] Briggs, L., Choi, C., Hu, W., Maas, L., Merk, B., Mochizuki, H., Monti, S., Morita, K., Del Nevo, A., Ohira, H., Petruzzi, A., Rtishchev, N., Sarathy, U., Shin, A., Shvetsov, I., Stempniewicz, M., Su, G., Sui, D., Truong, B., Vezzoni, B., Zanino, R., Zhang, Y., "EBR-II Passive Safety Demonstration Tests Benchmark Analyses – Phase 2", Proc. Int. Conf. NURETH-16, Chicago, IL, USA, August 30-September 4, 2015.

[4] Marchetti, M., Vezzoni, B., Gabrielli, F., Rineiski, A., "Neutron Deterministic Models of the EBR-II SHRT-45R Core Configuration", Transactions of the ANS, Vol. 111, Anaheim, CA, USA, November 9–13, 2014.

[5] Vezzoni, B., Marchetti, M., Gabrielli, F., Rineiski, A., "IAEA EBR-II Neutronics Benchmark: Impact of Modeling Options on KIT Results", to be presented at the Int. Conf. PHYSOR 2016, Sun Valley, Idaho, USA, May 1 – 5, 2016.

[6] Vezzoni, B., Marchetti, M., Andriolo, L., Gabrielli, F., Chen, X.-N., Matzerath Boccaccini, C., Rineiski, A., Maschek, W., Morita, K., Arima, T., "SIMMER/PARTISN Analyses of the EBR-II Shutdown Heat Removal Tests", to be presented at the Int. Conf. PHYSOR 2016, Sun Valley, Idaho, USA, May 1 – 5, 2016.

[7] Andriolo, L., Rineiski, A., Chen, X-N., Maschek, W., Gabrielli, F., Delage, F., Merle-Lucotte, E., "Extension of the SIMMER code for the assessment of core thermal expansion feedbacks for transient analysis", Proc. Int. Conf. IYNC 2014, Burgos, Spain, July 6-12, 2014.

[8] Andriolo, L., Rineiski, A., Vezzoni, B., Gabrielli, F., Chen, X.-N., Maschek, W., Delage, F., Merle Lucotte E., (2015). An innovative methodology for evaluating core thermal expansion feedbacks in transient analysis. International Congress on Advances in Nuclear Power Plants (ICAPP 2015), Nice, F, May 3-6, 2015 Proceedings on USB-Stick Paper 15292 Madison, Wis. : Omnipress, 2015.

[9] Andriolo, L., Impact of innovative sphere-pac fuels on safety performances of sodium cooled fast reactors. Fluids mechanics [physics.class-ph]. Universite Grenoble Alpes, 2015. English. <NNT: 2015GREAI067>.<tel-01257910> (https://tel.archives-ouvertes.fr/tel-01257910/document).

[10] Sumner T., Wei, T.Y.C., "Benchmark specifications and data requirements for EBR-II shutdown heat removal tests SHRT-17 and SHRT-45R," Tech. Rep. ANL-ARC-226, Argonne National Laboratory, May 2012.

[11] Marchetti, M., Gabrielli, F., Rineiski, A., Maschek, W., "The SIMMER/PARTISN Capability for Transient Analysis", Proc. Int. Conf. PHYSOR 2014, Kyoto, Japan, September 28-October 3, 2014.

[12] Chen, X.-N., Kiefhaber, E., "Asymptotic Solutions of Serial Radial Fuel Shuffling", Journal Energies, Special Issue "Sustainable Future of Nuclear Power", Asymptotic solutions of serial radial fuel shuffling. Energies, 8 (12), 13829-13845, 2015, doi:10.3390/en81212398

[13] Li, R., Wang, S., Rineiski, A., Zhang, D., Merle-Lucotte, E., "Transient Analyses for a Molten Salt Fast Reactor with Optimized Core Geometry", Journal Nuclear Engineering and Design, Vol. 292, pp. 164-176, October 2015, doi:10.1016/j.nucengdes.2015.06.011.

[14] Vezzoni, B., Gabrielli, F., Rineiski, A., (2015). Analyses of the performance of the ASTRID-like TRU burners in regional scenario studies. 21st International Conference and Exhibition: 'Nuclear Fuel Cycle for a Low-Carbon Future' (GLOBAL 2015), Paris, F, September 21-24, 2015 Proceedings., SFEN, Paris, F.

[15] Delage, F., Béjaoui, S., Lemehov, S., Somers, J., Maschek, W., . (2015). Outcomes of the FP-7 project PELGRIMM investigating pelletized ad sphere-packed oxide fuels for minor-actinides transmutation in a sodium fast reactor. 21st International Conference and Exhibition: 'Nuclear Fuel Cycle for a Low-Carbon Future' (GLOBAL 2015), Paris, F, September 21-24, 2015 Proceedings, SFEN, Paris, F.

[16] Maschek, W.; Andriolo, L.; Matzerath Boccaccini, C.; Delage, F.; Parisi, C.; Del Nuevo, A.; Abbate, G.; Schmitt, D. (2015). Simplified design and safety performance assessment of an advanced sphere-pac (U,Pu,MA)O2 SFR core. TopFuel 2015, Zürich, CH, September 13-17, 2015.

[17] Polidoro, F.; Flad, M.; Maschek, W. (2015). Preliminary evaluation of recriticality potential for ESFR core and conditions for its elimination. Nuclear Technology, 191, 246-253. doi:10.13182/NT14-97

[18] Vezzoni, B.; Gabrielli, F.; Rineiski, A.; Schwenk-Ferrero, A.; Romanello, V.; Maschek, W.; Fazio, C.; Salvatores, M. (2015). Plutonium and minor actinides incineration options using innovative Na-cooled fast reactors: Impacting on phasing-out and on-going fuel cycles. Progress in Nuclear Energy, 82, 58-63. doi:10.1016/j.pnucene.2014.07.034

[19] Chen, X. N.; Andriolo, L.; Rineiski, A.; Bubelis, E.; Mayer, G.; Bentivoglio, F. (2015). Extension and validation of SIMMER III code for gas cooled fast reactor. Annals of Nuclear Energy, 81, 320-331. doi:10.1016/j.anucene.2015.03.007

[20] Gabrielli, F.; Rineiski, A.; Vezzoni, B.; Maschek, W.; Fazio, C.; Salvatores, M. (2015). ASTRID-like fast reactor cores for burning plutonium and minor actinides. Energy Procedia, 71, 130-139. doi:10.1016/j.egypro.2014.11.863

[21] Li, R.; Chen, X. N.; Matzerath Boccaccini, C.; Rineiski, A.; Maschek, W. (2015). Study on severe accident scenarios: Pin failure possibility of MYRRHA-FASTEF critical core. Energy Procedia, 71, 14-21. doi:10.1016/j.egypro.2014.11.850

[22] Andriolo, L., Chen, X.-N., Matzerath Boccaccini, C., Rineiski, A., Maschek, W., Delage, F., Merle-Lucotte, E., "Preliminary Safety Analysis of Mixed Oxide Sphere-pac Driver Fuels in the CP-ESFR WH Core", Journal Energy Procedia, Vol. 71, pp. 149 – 158, May 2015.

[23] Zhang, D.; Rineiski, A.; Wang, C.; Guo, Z.; Xiao, Y.; Qiu, S. (2015). Development of a kinetic model for safety studies of liquid-fuel reactors. Progress in Nuclear Energy, 81, 104-112. doi:10.1016/j.pnucene.2015.01.011

[24] Chen, X. N.; Rineiski, A.; Gabrielli, F.; Andriolo, L.; Li, R.; Maschek, W. (2015). SIMMER-III parametric studies of fuel-steel mixing and radial mesh effects on power excursion on ESFR ULOF transients. International Congress on Advances in Nuclear Power Plants (ICAPP 2015), Nice, F, May 3-6, 2015 Proceedings on CD-ROM Paper 15033.

[25] Maschek, W.; Andriolo, L.; Boccaccini, C. M.; Delage, F.; Parisi, C.; Del Nevo, A.; Abbate, G.; Schmitt, D. (2015). Safety analyses for sodium-cooled fast reactors with pelletized and sphere-pac oxide fuels within the FP-7 European project PELGRIMM. International Congress on Advances in Nuclear Power Plants (ICAPP 2015), Nice, F, May 3-6, 2015.

[26] Li, R.; Chen, X. N.; Rineiski, A.; Moreau, V. (2015). Studies of fuel dispersion after pin failure: Analysis of assumed blockage accidents for the MYRRHA-FASTEF critical core. Annals of Nuclear Energy, 79, 31-42. doi:10.1016/j.anucene.2015.01.002

[27] Kuznetsov, V.; Fesenko, G.; Schwenk-Ferrero, A.; Andrianov, A.; Kuptsov, I. (2015). Innovative nuclear energy systems: state-of-the art survey on evaluation and aggregation judgment measures applied to performance comparison. Energies, 8, 3679-3719. doi:10.3390/en8053679

[28] Chen, X. N.; Li, R.; Rineiski, A.; Jäger, W. (2015). Macroscopic pin bundle model and its blockage simulations. Energy Conversion and Management, 91, 93-100. doi:10.1016/j.enconman.2014.11.053

[29] Chen, X. N. (2015). On LBE natural convection and its water experimental simulation. Progress in Nuclear Energy, 78, 372-379. doi:10.1016/j.pnucene.2014.03.001

[30] Chen, X.-N., Rineiski, A., Gabrielli, F., Andriolo, L., Vezzoni, B., Li, R., Maschek, W., Kiefhaber, E., "Fuel Steel Mixing and Radial Mesh Effects in Power Excursion", Journal Annals of Nuclear Energy, Vol. 90, pp. 26-31, April 2016.

[31] Li, R., Maschek, W., Matzerath Boccaccini, C., Marchetti, M., Kriventsev, V., Rineiski, A., "Bell-Plesset Instability Analysis for an Inward Centralized Sloshing", Journal Nuclear Engineering and Design, Vol. 297, pp. 312-319, February 2016.

[32] Kriventsev, V., "A Rule of Minimal Energy Dissipation in Modeling of Channel Flow Turbulence", Journal Numerical Heat Transfer Part B, Vol. 69, No. 2, pp. 85-95, February 2016.

Read more:

[33] Chen, X. N. (2015). On CFD methodology for simulations of whole reactor flow. 7th International Conference on Modelling and Simulation in Nuclear Science and Engineering (7ICMSNSE 2015), Ottawa, CDN, October 18-21, 2015.

[34] Chen, X. N.; Rineiski, A. (2015). Asymptotic radial fuel shuffling mode for accelerator driven transmuter. 17th International Conference on Emerging Nuclear Energy Systems (ICENES2015), Istanbul, TR, October 4-8, 2015.

[35] Gabrielli, F. (2015). Fuel for ADS: state-of-the-art, requirements, current and future programs. Technology and Components of Accelerator-Driven Systems 2nd International Workshop Proceedings, Nantes, F, May 21-24, 2013, 38-47, OECD/NEA, Paris, F.

[36] Guo, L.; Li, R.; Wang, S.; Flad, M.; Maschek, W.; Rineiski, A. (2015). Numerical investigation of SIMMMER code in a FCI premixing experiment. 17th International Conference on Emerging Nuclear Energy

Systems (ICENES 2015), Antalya, TR, May 10-14, 2015.

[37] Kriventsev, V.; Rineiski, A.; Pfrang, W.; Perez-Martin, S.; Mikityuk, K.; Khvostov, G. (2015). Benchmark on behavior of MOX fuel pin under irradiation at nominal power in sodium fast reactor. TopFuel 2015, Zürich, CH, September 13-17, 2015.

[38] Li, R.; Maschek, W.; Kriventsev, V.; Guo, L.; Rineiski, A. (2015). Bell-Plesset instability analysis for an inward centralize sloshing. 17th International Conference on Emerging Nuclear Energy Systems (ICENES 2015), Antalya, TR, May 10-14, 2015.

[39] Schwenk-Ferrero, A.; Tromm, T. W.; Bohnstedt, A. (2015). R&D activities in German nuclear sector. 23rd WiN Global Annual Conference: Women in Nuclear Meet Atoms for Peace, Wien, A, August 24-28, 2015.

[40] Vezzoni, B.; Gabrielli, F.; Rineiski, A.; Schwenk-Ferrero, A.; Andriolo, L.; Maschek, W. (2015). Innovative TRU burners and fuel cycles options for phase-out and regional scenarios. Actinide and Fission Product Partitioning and Transmutation : 13th Information Exchange Meeting, Seoul, Korea, September 23-26, 2014 Proceedings., 104-114, OECD/NEA, Paris, F.

Group: Accident Analysis

Analysis of Design Basis and Severe Accidents

Alexei Miassoedov, Giancarlo Albrecht, Thomas Cron, Philipp Dietrich, Beatrix Fluhrer, Jerzy Foit, Stephan Gabriel, Xiaoyang Gaus-Liu, Frank Kretzschmar, Markus Schwall, René Stängle, Mike Vervoortz, Thomas Wenz

Introduction

The analysis of severe accidents in LWRs at the IKET group UNA is focused on the in- and ex-vessel core melt behavior. The overall objective is to investigate the core melt scenarios from the beginning of core degradation to melt formation and relocation in the vessel, possible melt dispersion to the reactor cavity and to the containment, and finally corium concrete interaction and corium coolability in the reactor cavity.

The experimental platform includes three experimental facilities:

- LIVE to investigate the melt pool behavior in the RPV lower head;
- DISCO to study the melt dispersion to the reactor cavity and direct containment heating;
- MOCKA to study molten corium concrete interaction.

The results of the experiments are being used for the development and validation of codes applied for safety assessment and planning of accident mitigation concepts, such as MELCOR and ASTEC. The strong coupling between the experiments and analytical activities contribute to a better understanding of the core melt sequences and thus improve safety of existing reactors by severe accident mitigation measures and by safety installations where required.

The understanding of major processes for the assessment of the plant response and behavior under design basis or beyond design basis situations still have to be further developed. For this purpose, experimental investigations are carried out in the WENKA and COSMOS facilities addressing the thermal-hydraulics and physico-chemical phenomena during postulated design and beyond design transients and accidents.

LIVE experiments

The main objective of the LIVE program is to study the late in-vessel core melt behavior and core debris coolability both experimentally in large scale 2D and 3D geometry and in supporting separate-effects tests, and analytically using CFD codes in order to provide a reasonable estimate of the remaining uncertainty band under the aspect of safety assessment. The LIVE-3D test facility allows the investigation of a melt pool in the lower plenum of a RPV in 3D geometry with simulated internal heat generation. Other test facilities had only a 2D geometry or were performed without heating of the melt. The main part of the LIVE-3D test facility is a 1:5 scaled semi-spherical lower head of the typical pressurized water reactor.

The diameter of the test vessel is 1 meter. The top area of the vessel is covered with an insulation lid. The test vessel is enclosed in a cooling vessel to simulate the external cooling. The melt is prepared in an external heating furnace designed to generate 220 l of the simulant melt. The volumetric decay heat is simulated by 6 spirally formed heating elements providing a maximum power of about 28 kW. To investigate both the transient and the steady state behavior of the simulated corium melt, an extensive instrumentation of the test vessel is realized. The temperatures of the vessel wall inner surface and outer surface are measured at 5 latitudes and 4 locations at each latitude. Heat flux distribution through the vessel wall can be calculated based on these temperatures. Additionally, 80 thermocouples are positioned within the vessel to measure the temperature distribution in the melt pool and in the crust.

In 2015 the results of single layer LIVE tests were systematically analyzed. By comparing the published data from other experimental facilities the LIVE test results can contribute to a better understanding of remaining issues regarding the heat transfer behavior in the melt pool. These issues are the influence of crust formation on molten pool heat transfer, the 2D effect and the heat flux distribution under different boundary condition.

Effect of crust formation

Large discrepancy in Nu~Ra correlations (Nusselt Na over Rayleigh Ra number) of upward and downward heat transfer were observed depending on whether a crust layer exists on the cooling boundary. For the upward heat transfer, higher Nu_{up} number is obtained in crust-forming liquid. LIVE experiments with nitrate salt as simulant material has shown similar behavior in comparison with water experiments, that both the upward and downward Nu number in case of crust formation are higher than the Nu number without crust formation. In Fig. 1 the Nu_{up} are shown, in which the Nu numbers with crust formation are marked with red color. In the reactor case, a crust layer should be formed in an oxide melt pool. With crust formation the melt/crust interface is ideal isothermal. Without crust formation, the temperature at the interface, namely the wall inner surface temperature is not uniform and is higher than the wall outer surface temperature; even the wall outer surface can be maintained isotherm with a cooling medium during phase transition. In the experiments without crust formation, the heat flux is underestimated when the wall outer surface temperature is assumed as the interface temperature for the heat flux distribution. In addition, the uneven crust surface promotes the local convective heat transfer.

Moreover, the heat flux distribution on the curved side wall shows different character as the heat flux without crust formation. As shown in Fig. 2, in salt tests with crust formation the heat flux profile through the vessel wall was flat at the vessel bottom and increases with steady slope towards the melt upper surface. It is supposed that the heat transfer inside the crust layer due to the temperature gradient is responsible for this effect. A crust layer can homogenize the vertical heat flux difference at the bottom and reduce the highest heat flux at the melt surface.

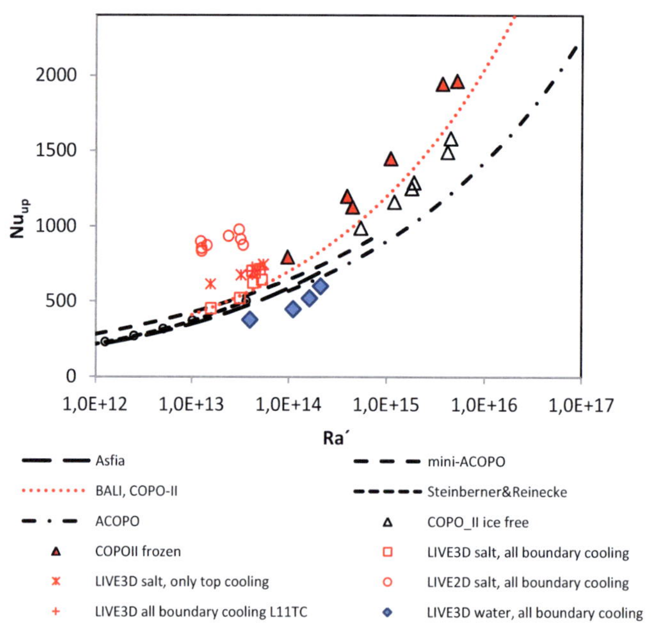

Fig. 1: Upward Nu numbers in LIVE tests with (red) and without crust formation

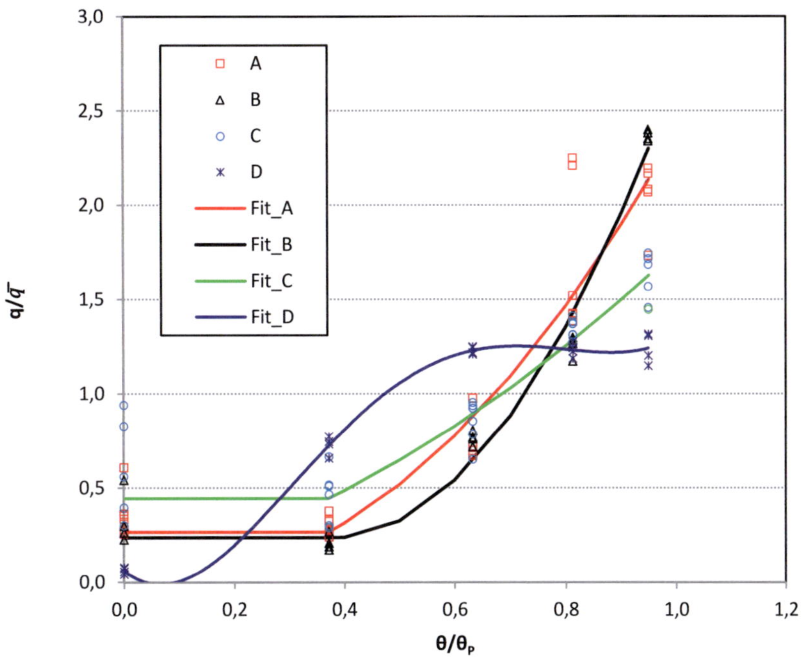

A: salt test, top insulation, external subcooling
B: salt test, top insulation, external boiling
C: salt test,all boundary subcooling
D: water test, all boundary subcooling

Fig. 2: Normalized heat flux distribution through vessel wall in LIVE 3D tests

<u>Effect of 2D geometry</u>

There is no boundary restriction of the turbulent flow in 2D geometry since the wave length of the thermal instabilities is more than 10 times smaller than the width of a 2D cavity. However, the aspect ratio describing the heat transfer area between upward and sidewall surfaces, here defined as $A_{HT} = A_{\_up}/A_{\_dn}$, shows the difference of up/down heat transfer area splitting between 2D and 3D geometry: A_{HT_3D} is twice of A_{HT_2D}, indicating that larger upper surface area in 3D pool is able to remove larger fraction of heat and thus mitigate the intensity of upward heat flux and the heat flux tapering on the side wall.

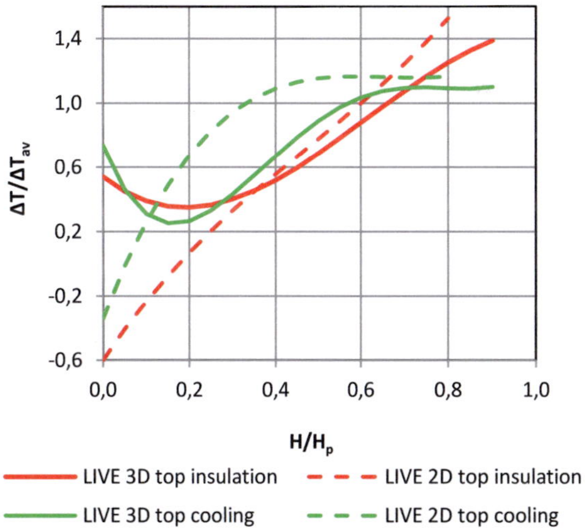

Fig. 3: Comparison of melt temperature distribution in LIVE3D and LIVE 2D

The comparison of the normalized temperature distribution in Fig. 3 and the normalized heat flux distribution in Fig. 4 demonstrate the effect of heat transfer in 2D geometry. The influence of the 2D effect should be given special concern by using the general results originated from 2D experiments.

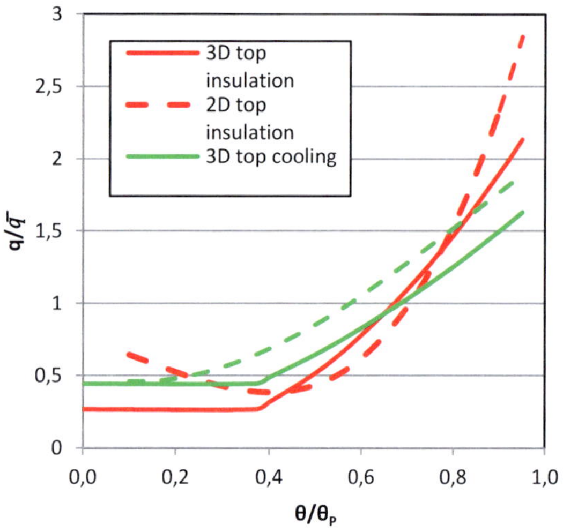

Fig. 4:. Comparison normalized heat flux distribution in LIVE3D and LIVE 2D

MOCKA experiments

The MOCKA facility is designed to investigate the corium/concrete interaction in an anticipated core melt accident in LWRs, after the metal melt is layered beneath the oxide melt. The experimental focus is on the cavity formation in the basemat and the risk of a long-term basemat penetration by the metallic part of the melt.

In MOCKA experiments cylindrical concrete crucibles with an inner diameter of 25 cm are used. Both the sidewall and the bottom are instrumented with Type K thermocouple assemblies to determine the concrete erosion as a function of time. The initial melt consists of 42 kg Fe together with 4 kg Zr, overlaid by 68 kg oxide melt (initially 56 wt.% Al_2O_3, 44 wt.% CaO). The initial height of the metal melt is about 13 cm. The melt temperature at start of interaction was approximately 2000 °C. The initial temperature was estimated from the reaction enthalpy of the used thermite taking into account the temperature measurements from the former BETA tests. The CaO admixture lowers the solidus temperature and the viscosity of the oxide melt. The resulting solidus temperature of approx. 1350 °C is sufficiently low to prevent a formation of an initial crust at the oxide/concrete interface. The internal heat generation in the oxide phase is simulated by a succession of additions of pure thermite and Zr metal to the melt from the top being the first of a kind heating method realized for high temperature melts worldwide. The heat generated by the exothermal oxidation reactions of the continuously added Zr is deposited in the oxide phase. Due to density-driven phase segregation the metal melt at the bottom of the crucible is fed by the enthalpy of the Fe melt which is generated in the oxide phase by the thermite reaction of the added thermite. Approximately 75 % of the heating power was deposited in the oxide phase and 25 % in the iron melt. In this way a prototypic heating of both melt phases can be achieved.

The heating method used in MOCKA allows investigation of MCCI process with reinforced concrete. The rebars in the concrete elevate the melt/concrete interface temperature up to the melting temperature of the reinforcing steel, i.e. 1528 °C. This should result in higher melt pool temperatures than during MCCI with concrete without reinforcement for which a concrete decomposition temperature of approximately 1300 °C was estimated. After a fast initial decrease in melt temperature, the temperature of the oxide melt cluster around 1600 °C in MOCKA 7.1 test with rebars, Fig. 5. Surprisingly, the same long-term temperature was also found in MOCKA 6.3 test without rebars, Fig. 6. As the thermal properties of the pure concrete and that of the reinforced concrete do not differ much and, in addition, the same internal heating power was generated in all experiments under consideration, the former estimated decomposition temperature of the pure concrete cannot be used to prescribe the temperature boundary condition for the MCCI process. The melt/ concrete interface temperature relevant to the MCCI process must be higher than the commonly used concrete decomposition temperature.

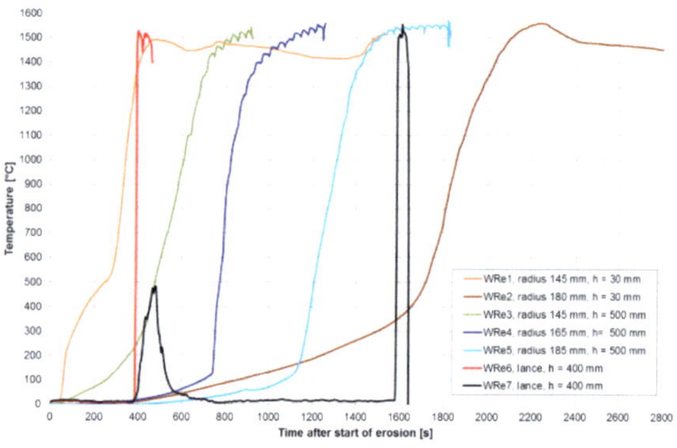

Fig. 5: Melt temperatures at different positions as a function of time in MOCKA 7.1 test

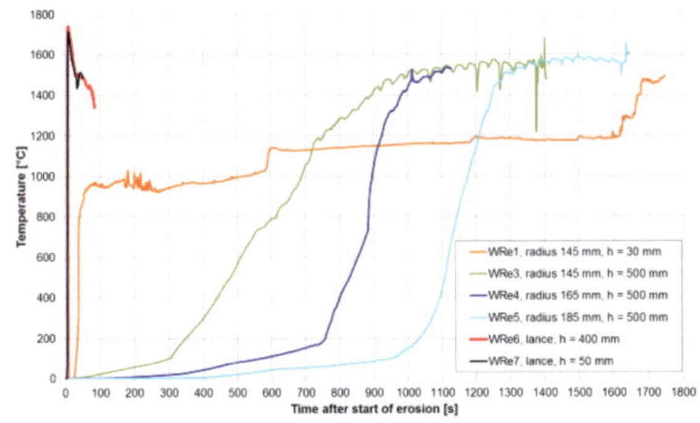

Fig. 6: Melt temperatures at different positions as a function of time in MOCKA 6.3 test

Recently performed MOCKA experiments with siliceous concrete gave rise to a similar conclusion. An experimental program to investigate the concrete decomposition process at high temperatures is underway. In contrast to the fairly isotropic concrete ablation in the CCI 2 experiment a highly pronounced lateral ablation was observed in all MOCKA tests using LCS concrete with and without rebars. The corresponding lateral/axial ratios of the ablation depth are approximately 3 (approximately 1 in CCI 2 test performed at ANL). The knowledge of relation between the axial and lateral basemat erosion is important in evaluation of the consequences of a severe reactor accident. However, the extrapolation of the obtained results to reactor conditions is difficult.

COSMOS experiments

The objective of the COSMOS (Critical-heat-flux On Smooth and Modified Surfaces) experiments is to study the influence of safety-relevant components of light water reactors (fuel assemblies, grid spacers etc.) on heat removal, boiling crisis, pressure loss and other important issues. The focus is on how these parameters are influenced by the material properties of the structures. New measurement technologies like fiber optic void sensors will be used to provide qualified data for the development and validation of physical correlations used in the systems codes and for the validation of detailed CFD codes in the long-term. The focus shall be on the use of up to date and innovative measurement techniques (e.g. LDA, PIV, high-speed cameras) of high spatial and temporal resolution and the possibility of their use in applications involving high temperatures and pressures.

The water loop for flow boiling heat transfer experiments at low pressure COSMOS-L was constructed and put into operation. A comprehensive series of two-phase flow experiments was successfully completed. The influence of the rod surface structure on critical heat flux (CHF) for flow boiling of water was investigated for Zircaloy tubes in a vertical annular test section. Though only a small influence of surface structures on critical heat flux was observed for the pressure of 120 kPa in the COSMOS-L test section the increase of CHF value is expected to be more pronounced (up to 30%) at higher pressure for certain clad surface modifications, like oxide layers, formed on the clad surface of operating fuel rods.

To quantify these phenomena for high pressures with an objective of increasing the safety margin to the start of the boiling crisis in prototypic conditions, the COSMOS-H experimental facility is being constructed. The facility consists of a high-pressure water loop and two cooling circuits and is designed for investigations of the CHF under non-scaled thermohydraulic conditions. The test section of the facility is approximately 3.7 m high and includes an outer pressure tube, which encloses the test fuel rod bundle guided by a circular or rectangular separation tube. The test section consists of several segments including those that are specially designed for the visual analysis of the flow. The modular construction of the section allows different vertical positioning of these special segments depending on the experimental conditions and test requirements (Fig. 7). To test and approve the test section design a single cladding tube will be used in the first experiments. In the future tests larger fuel rod bundles of 2x2, 3x3 and hexagonal arrangements will be included in the study.

The low pressure loop COSMOS-L was used in 2015 mostly for testing the high pressure COSMOS-H concept, material selection and instrumentation for the test section of the COSMOS-H facility.

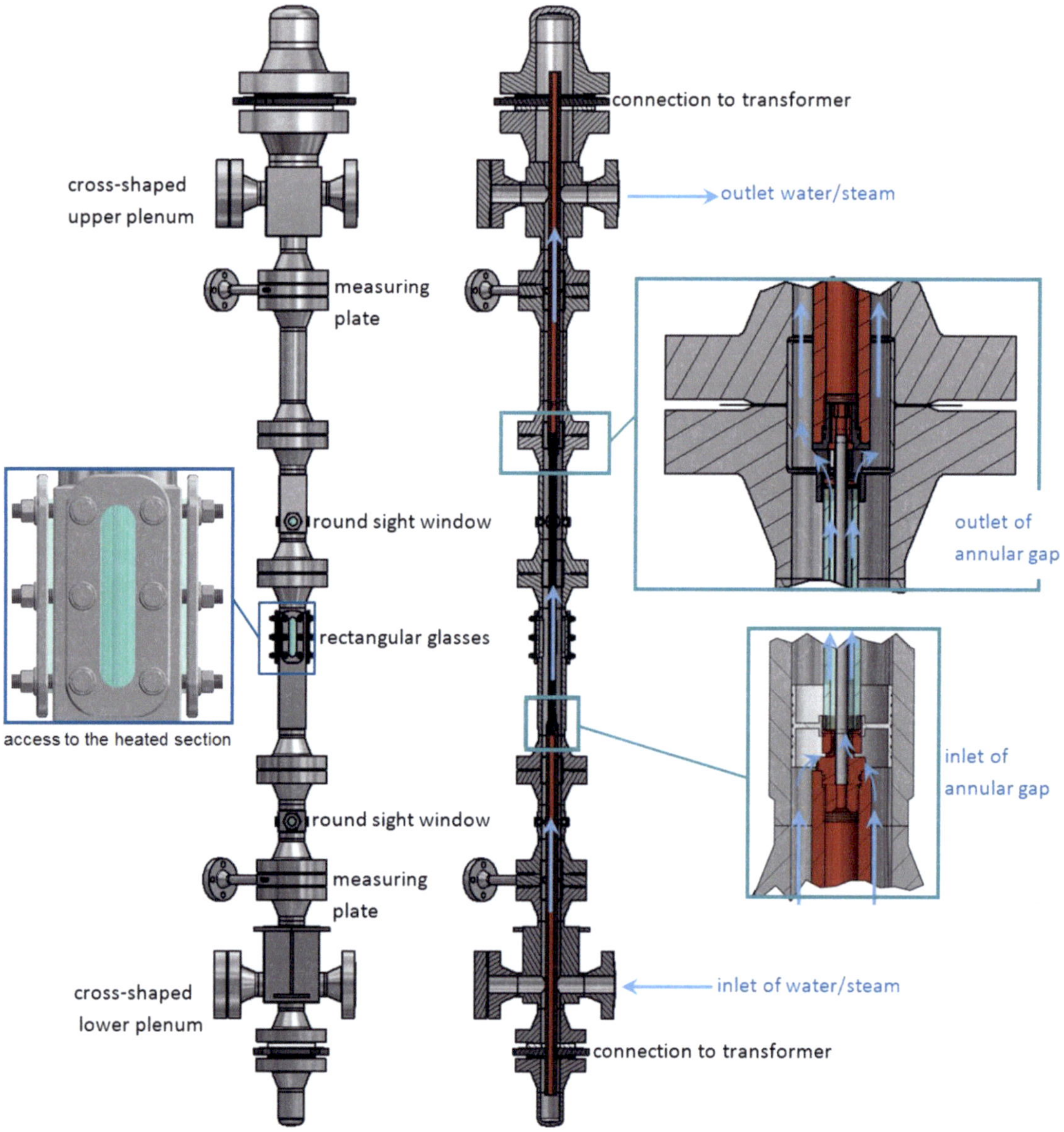

Fig. 7: Scheme of the COSMOS-H test section

WENKA experiments

The WENKA channel (Water Entrainment channel Karlsruhe) serves the investigation of counter-current stratified two-phase flows. These complex flow phenomena are relevant in accident scenarios of pressurized water reactors like the reflux condensation mode after a Loss Of Coolant Accident (LOCA). The objective of the experimental work is the provision of detailed datasets for development and validation of turbulence and phase interaction models for the CFD codes for which local measurements of various flow parameters with high spatial resolution are required. The WENKA channel has a modular construction concept and operates under ambient conditions. Water and air flow rates can be controlled in a wide range. The current configuration allows the observation of different flow regimes, such as supercritical flow, supercritical flow with droplet entrainment, partially reversed flow and fully reversed flow. These flow regimes include flow phenomena like hydraulic jumps, bubble and droplet entrainment and propagation of capillary and gravity waves.

For the measurements advanced optical methods are used such as intrusive devices. The velocity and velocity-fluctuation profiles are measured by particle image velocimetry (PIV). For measurement in water fluorescent tracer particles are used to suppress reflections at the liquid surface. To provide high-resolved void fraction data a special method was developed. The method uses sequences of high speed images to calculate a 2D distribution of the volumetric void fraction. After the image post-processing the probability of existence of a two-phase region is calculated by averaging over a period of 16 s; the void fraction is then calculated by integration of the intensity values. The validation of the method was performed by comparing its results to the electric needle probe data, demonstrating a measurement uncertainty of 8.6 %. Using these and other instruments an extensive database containing high-speed movies, flow maps, turbulence and velocity data including droplet mass flow has been obtained within the last three years. The measurement data can be directly compared with the results of CFD simulations.

Application of the MELCOR code

The UNA group participates in the WASA-BOSS (Weiterentwicklung und Anwendung von Severe Accident Codes – Bewertung und Optimierung von Störfallmaßnahmen) project. This project is supported by the German Federal Ministry of Education and Research (BMBF). Task of the UNA group is evaluating and improving the capability of the severe accident code MELCOR to calculate the behaviour of molten core material, which relocated to the lower plenum during a hypothetical severe accident in a nuclear power plant. In order to improve the prediction of such molten core material several new models have been developed. For example, the Phase-Change Effective Convectivity Model (PECM) developed at the Royal Institute of Technology (KTH) is able to provide a more detailed and precise description of the molten pool behaviour in the lower head in the same calculation time as MELCOR.

The applied approach consists of coupling new models such as the PECM to the system code MELCOR. Therefore, the program DINAMO (Direct Interface for Adding Models) was developed. This program allows the calculations of specific phenomena of a severe accident by external models in MELCOR without the necessity of complex modifications of the MELCOR source code. In order to improve the prediction of molten core material in the lower plenum the PECM, which has been implemented into the CFD software OpenFOAM, is coupled to MELCOR via the coupling tool DINAMO. First coupled simulations on the experiment LIVE-L1 showed that the coupled system can improve the prediction of a severe accident in MELCOR.

Based on these results, additional LIVE-tests have been simulated to further validate the coupled MELCOR-PECM system as well as the participating codes. In LIVE-L10 and LIVE-L11 the influence of different external cooling conditions on core melt behaviour in the lower plenum is investigated. In LIVE-L10 a constant coolant support was established to realize sub-cooled cooling conditions with a non-boiling coolant. A coolant support based on the current water level is used in the LIVE-L11 test to realize boiling of the coolant. Apart from the cooling conditions the two experiments show the same setup. In Fig. 8 the vertical temperature profiles in the coupled MELCOR-PECM system for the two experiments with a volumetric heating power of 21 kW compared to the experimental findings and a standalone MELCOR calculation is shown.

The mean temperature predicted by the coupled system matches the mean temperature measured in the experiments. For both experiments in MELCOR this temperature is overestimated. In addition, standalone MELCOR, which cannot resolve the temperature distribution, predicts a significant temperature increase in the molten material in LIVE-L11 compared to LIVE-L10, which was not observed in the experiments. The independence of the behaviour of the molten material in the lower plenum from the external cooling conditions found in the experiments is predicted correctly by the coupled MELCOR-PECM system.

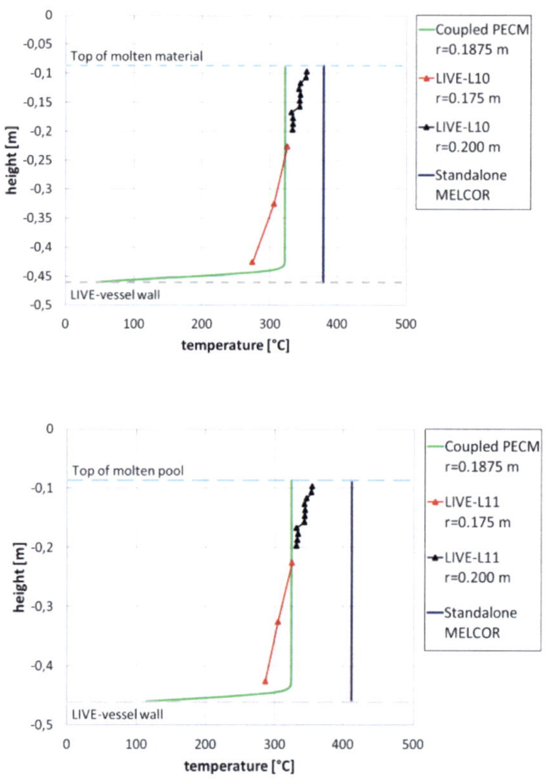

Fig. 8: Vertical temperature profiles for LIVE-L10 (top) and LIVE-L11 (bottom) with 21 kW volumetric heating

The coupled system was further tested by simulating an entire nuclear power plant under severe accident conditions. Therefore, a MELCOR input deck for a generic western pressurized water reactor shown in Fig. 9 was developed.

In flow path 228 a leak opens after 500 s. Apart from the water provided by hydro-accumulators, no coolant is injected into the primary circuit to mitigate the severe accident. In order to provide external cooling for the molten core material in the lower head the reactor cavity (control volume 700 in Fig. 9) is flooded with water. In Fig. 10 the pressure and water level in the reactor pressure vessel in the coupled simulation is compared to the results of a standalone MELCOR calculation.

Until the end of the coupled simulation with the PECM, which is induced by the failure of the lower head, the pressure and water level are in a very good agreement. However, the failure of the lower head is calculated about 1 hour earlier compared to the standalone MELCOR calculations. Due to the total simulation time of about 5 hours this significantly affects other potential severe accident management strategies. Based on the more accurate prediction of the LIVE tests the result of the coupled simulation is assumed to be more realistic. This first coupled simulation of molten material in the lower head shows the benefits of the coupling methodology.

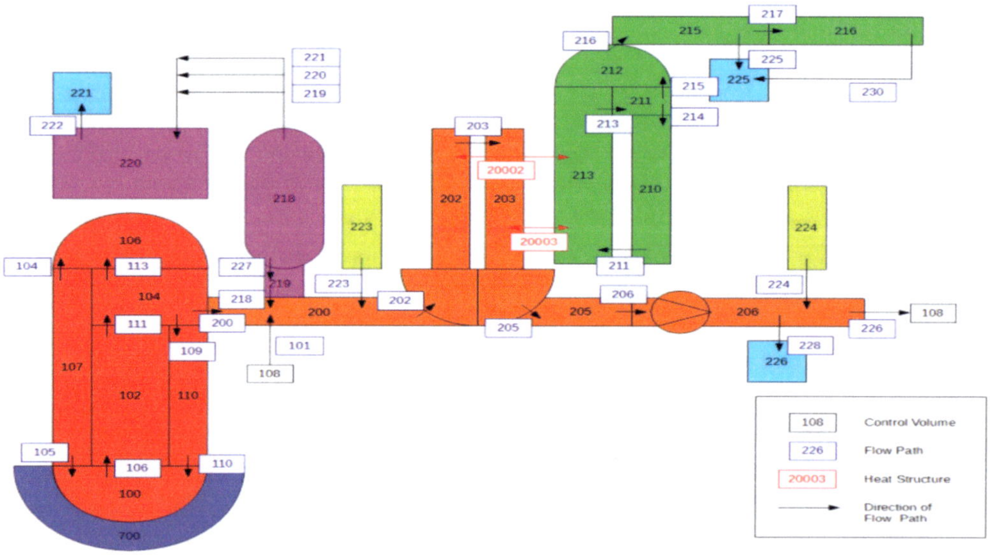

Fig. 9: Scheme of the developed MELCOR input deck for a generic western pressurized water reactor

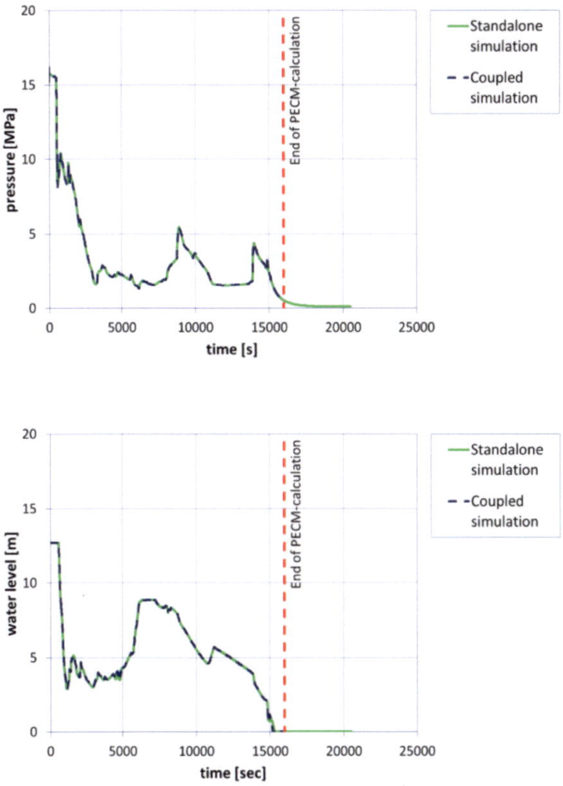

Fig. 10: Pressure (top) and water level (bottom) in the reactor pressure vessel in the coupled and the standalone simulation

In addition to the external cooling of the molten material in the lower head, water can be injected into the reactor pressure vessel to cool the molten material from the top. This cooling configuration was investigated in the LIVE-L7v test. In order to further validate the PECM, this experiment was simulated. The simulation results showed significant deviations to the experimental findings. The PECM overestimates the mean temperature in the molten material and cannot resolve the stratified region in the bottom part of the molten pool. Therefore, the PECM was modified. The developed stratified PECM (S-PECM) replaces the PECM-methodology in the stratified region by an "Effective Conductivity" approach. Fig. 11 shows the comparison of the vertical temperature profile with 29 kW of the S-PECM with the traditional PECM and the experimental results from LIVE-L7v.

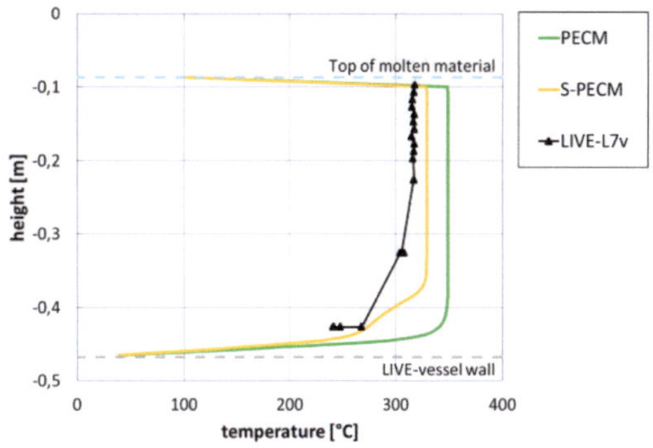

Fig. 11: Vertical temperature profiles at r = 0.175 m

The S-PECM shows deviations compared to the experiment, however, it significantly better predicts the temperature profile than the traditional PECM. In the lower region also the stratification is resolved by the S-PECM. This modified model allows the improved simulation of molten core material in the lower plenum, which is cooled from the top and through the lower head wall.

References

[1] Bakalov, I.; Spengler, C.; Foit, J. (2015). Analysis of MOCKA experiments for MCCI with stratified oxide and metal melts and internal heat generation by enthalpy feeding using the MEDICIS code. 7th Conference on Severe Accident Research (ERMSAR 2015), Marseille, F, March 24-26, 2015. Proceedings, Paper 062.

[2] Dietrich, P.; Gaus-Liu, X.; Miassoedov, A.; Kretzschmar, F.; Class, A. (2015). LIVE experiments on the in-vessel melt pool behaviour with external cooling conditions and the simulation results with a coupled MELCOR-PECM approach. 16th International Topical Meeting in Nuclear Reactor Thermal Hydraulics (NURETH-16), Chicago, Ill., August 30 - September 4, 2015 Proceedings, 5658-5671, ANS, LaGrange Park, Ill.

[3] Dietrich, P.; Kretzschmar, F.; Miassoedov, A.; Class, A. (2015). WASA-BOSS: Expansion of the model-basis in MELCOR. 46.Jahrestagung Kerntechnik, Berlin, 5.-7.Mai 2015, CD-ROM, INFORUM GmbH, Berlin.

[4] Dietrich, P.; Kretzschmar, F.; Miassoedov, A.; Class, A.; Villanueva, W.; Bechta, S. (2015). Coupling of MELCOR with the PECM for improved modelling of a core melt in the lower plenum. 23rd International Conference on Nuclear Engineering (ICONE23), Chiba, J, May 17-21, 2015. Proceedings on DVD, ICONE23-1593, The Japan Society of Mechanical Engineers, Tokyo.

[5] Dietrich, P.; Kretzschmar, F.; Miassoedov, A.; Class, A.; Villanueva, W.; Bechta, S. (2015). Extension of the MELCOR code for analysis of late in-vessel phase of a severe accident. Proceedings of the 5th International Youth Conference on Energy (IYCE 2015), Pisa, I, May 27-30, 2015, Article nr 7180747, IEEE, Piscataway, NJ. doi:10.1109/IYCE.2015.7180747

[6] Fischer, M.; Bechta, S. V.; Bezlepkin, V. V.; Hamazaki, R.; Miassoedov, A. (2015). Core melt stabilization concepts for existing and future LWRs and associated R&D needs. 16th International Topical Meeting in Nuclear Reactor Thermal Hydraulics (NURETH-16), Chicago, Ill, August 30 - September 4, 2015 Proceedings, 7578-7592, ANS, LaGrange Park, Ill.

[7] Foit, J. J. (2015). From experiment to reactor: A MCCI scaling issue. 23rd International Conference on Nuclear Engineering (ICONE23), Chiba, J, May 17-21, 2015.

[8] Foit, J. J. (2015). MCCI on LCS concrete with and without rebars. 7th Conference on Severe Accident Research (ERMSAR 2015), Marseille, F, March 24-26, 2015. Proceedings, Paper 008.

[9] Journeau, C.; Liao, Y.; Zhang, H.; Miassoedov, A.; Tian, W.; Bo, K.; Gaus-Liu, X. (2015). Access to large infrastructures for severe accidents in Europe and in China: The ALISA project. 7th Conference on Severe Accident Research (ERMSAR 2015), Marseille, F, March 24-26, 2015. Proceedings, Paper 24.

[10] Miassoedov, A.; Albrecht, G.; Foit, J.; Gaus-Liu, X.; Tromm, W. (2015). LWR severe accident research at the Karlsruhe Institute of Technology. 23rd ABCM International Congress of Mechanical Engineering (COBEM 2015), Rio de Janeiro, BR, December 6-11, 2015.

[11] Miassoedov, A.; Haas, C.; Gabriel, S.; Albrecht, G. (2015). Study of the critical heat flux on zircaloy cladding surfaces in the COSMOS facilities. Transactions of the American Nuclear Society, 113, 1478-1481.

[12] Miassoedov, A.; Journeau, C.; Bechta, S.; Hozer, Z.; Manara, D.; Bottomley, D.; Kiselova, M.; Langrock, G. (2015). Severe accident facilities for European safety targets. The SAFEST project. 16th International Topical Meeting in Nuclear Reactor Thermal Hydraulics (NURETH-16), Chicago, Ill., August 30 - September 4, 2015 Proceedings, 4604-4616, ANS, LaGrange Park, Ill.

[13] Miassoedov, A.; Journeau, C.; Bechta, S.; Hozer, Z.; Manara, D.; Kiselova, M.; Langrock, G.; Schyns, M. (2015). Severe accident facilities for European safety targets. The SAFEST project. 7th Conference on Severe Accident Research (ERMSAR 2015), Marseille, F, March 24-26, 2015. Proceedings, Paper 21.

[14] Miassoedov, A.; Fluhrer, B.; Journeau, C.; Bechta, S.; Hozer, Z.; Manara, D.; Bottomley, P. D. W. I.; Kiselova, M.; Schyns, M. (2015). The SAFEST project: towards a pan-European laboratory on corium experimental R&D. 23rd International Conference on Nuclear Engineering (ICONE 23), Chiba, J, May 17-21, 2015.

[15] Stefanova, A.; Gencheva, R.; Groudev, P.; Cranga, M.; Spengler, C.; Ivanov, I.; Foit, J.; Kostka, P. (2015). Investigation of molten core-concrete interaction reactor benchmark test for VVER 1000. Annals of Nuclear Energy, 75, 742-751. doi:10.1016/j.anucene.2014.07.021

[16] Van Dorsselaere, J. P.; Auvinen, A.; Beraha, D.; Chatelard, P.; Herranz, L. E.; Journeau, C.; Klein-Hessling, W.; Kljenak, I.; Miassoedov, A.; Paci, S.; Zeyen, R. (2015). Recent severe accident research synthesis of the major outcomes from the SARNET network. Nuclear Engineering and Design, 291, 19-34. doi:10.1016/j.nucengdes.2015.03.02

Group: KArlsruhe Liquid metal LAboratory

Liquid metal thermal hydraulics for energy technologies

Tobias Geißler, Klarissa Niedermeier, Julio Pacio, Thomas Wetzel „and the KALLA team“

Introduction

For the Karlsruhe Liquid Metal Laboratory (KALLA), 2015 has again been an exciting year with some highly interesting and motivating research results and steps forward. The general segments of our research in liquid metal (LM) technology have been maintained:

- Experiments and modeling of liquid metals as liquid phase medium in a bubble column chemical reactor, in particular for direct pyrolysis of methane
- Investigation of LMs as efficient heat transfer fluids for concentrated solar power (CSP) systems
- Experiments for cooling of core components of Accelerator Driven Systems (ADS)

Some highlights of our research will be outlined in the following sections. One major success of the whole laboratory was the 3-year evaluation of the HELMHOLTZ-Alliance LIMTECH, where KALLA is one of the founding partners. An international expert panel has evaluated the work being done in the frame of this research alliance since 2013, with a very positive end encouraging judgement. At KALLA, 4 PhD students and one senior researcher are performing work in LIMTECH.

Hydrogen/carbon production by methane cracking

The project tries to shed light on the possibility to produce gaseous hydrogen and solid carbon from methane, without producing CO_2 as a side product. It was embedded in a cooperation with the Institute for Advanced Sustainability Studies (IASS) in Potsdam and still is part of the Helmholtz Alliance for Liquid Metal Technologies (LIMTECH). The main drawback of former attempts to realize a thermal methane cracking process – mainly in tubular reactors – is the deposition of solid carbon layers on the heated reactor walls, subsequently leading to reactor blockage. Removing these carbon layers poses a severe problem, which has so far prevented continuous large scale industrial application of this process. Within this project, an alternative approach for the continuous cracking of methane was developed and investigated in the HELiS laboratory scale reactor at KIT, KALLA. The idea is the utilization of liquid metal as a heat transfer fluid, which is chemically stable and applicable in a bubble column reactor at temperatures above 1200 °C. In the investigated process, methane gas is injected via a 0.5 mm single orifice into a liquid metal bubble column reactor, made of quartz glass with an inner diameter of 40.6 mm and a total length of 1300 mm. The methane decomposes inside of the formed bubbles which are rising up in the reactor.

The bubble opens at the upper interface of the liquid metal, releasing the produced carbon, which is floating on the liquid metal surface and hydrogen but also the remaining methane gas and the formed gaseous intermediates. The advantages of this process are the continuous formation of new small scale reaction volumes represented by the injected bubbles and the natural carbon separation due to the density difference between the liquid metal and the carbon. During 2015, several experimental campaigns with and without the application of a packed bed were conducted and investigations in terms of temperature, flow rate and geometry influence were performed. The resulting hydrogen yield as a function of the predominant reference temperature under different conditions is shown in Fig. 1.

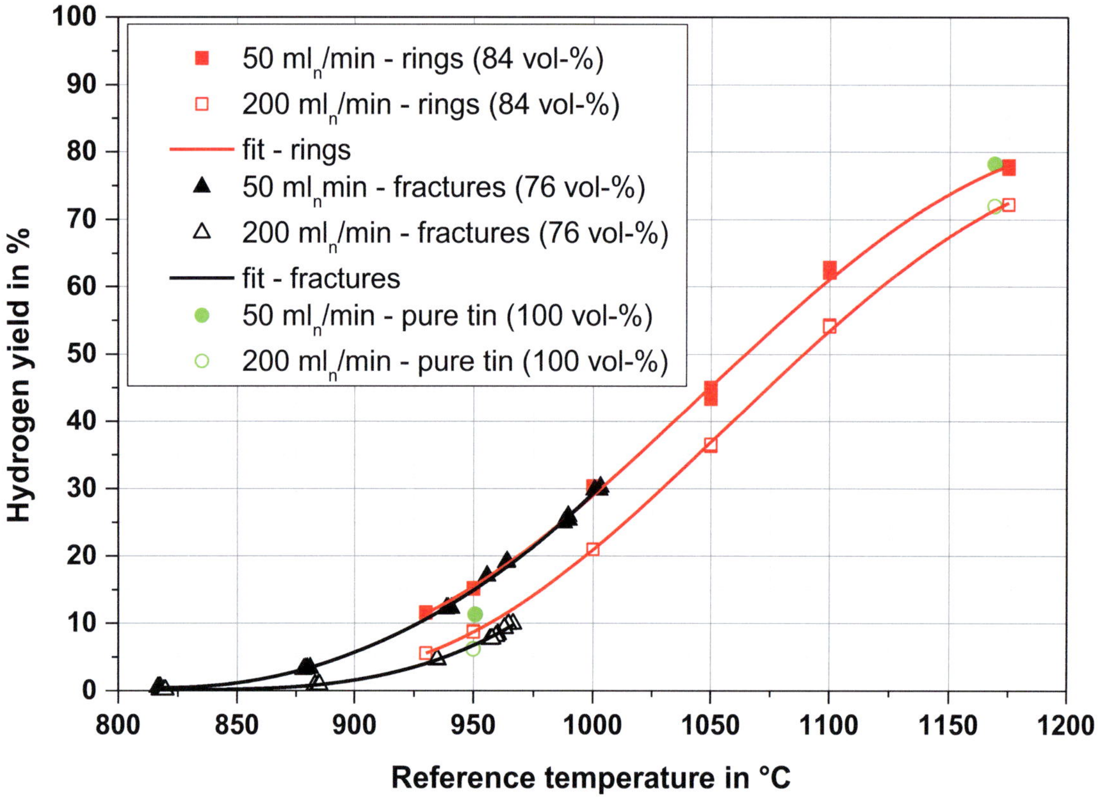

Fig. 1: Hydrogen yield as a function of reference temperature at 50 ml_n/min and 200 ml_n/min methane volume flow rate for two different packed bed geometries in comparison with pure tin experiments.

By installing a new 8 kW electric furnace with three independent heating zones, the maximum operating temperature has been enhanced to 1175 °C with a resulting hydrogen yield of maximum 78 % at 50 ml_n/min methane volume flow rate. This temperature has not been realized in a liquid metal reactor for methane cracking until now. Also the absence of considerable amounts of intermediates is promising for future process development, as the selectivity for hydrogen production is almost one. In general, the hydrogen yield increases by increasing the liquid metal temperature and the gas residence time in the reactor. No significant impacts of the investigated packed bed geometries on the hydrogen yield compared with pure tin experiments were detected. In contrast to investigations in tubular reactors, no carbon clogging occurred and the produced carbon accumulated as a loose powder on the liquid metal surface in the upper part of the reactor, shown in Fig. 2.

The results have found widespread echo in the scientific as well as in the interested general public after KIT issued a press release on the successful steps in fall 2015. Several dozen international journals and newspapers reported about KALLA and the methane cracking, including Frankfurter Allgemeine, The Economist, Deutschlandfunk, Wirtschaftswoche, to name just a few.

Fig. 2: Carbon powder accumulation in the upper part of the reactor after 14 days of operation.

Concentrated Solar Power (CSP)

Liquid metals have been proposed in the past as high temperature heat transfer media in concentrating solar power (CSP) systems and have regained much attention in recent years, not at last due to well recognized work in this field by KALLA researchers. In 2015, as a major step towards liquid metal thermal hydraulics tests under concentrated sunlight conditions, the SOMMER (**SO**lar furnace arrangement with **M**olten-**ME**tal-cooled **R**eceiver) facility has been completed. Calibration and verification measurements by a specialized external company have been accomplished in November.

Lead-bismuth-eutectic (LBE) mixture has been selected for the receiver first test loop. The thermal receiver of the SOMMER loop will be located in the focal point of the parabolic concentrator, the latter with a square aperture area of 16 m². The mirror has a focal length of 2.4 m and consists of nine glass facets with 3 mrad expected total slope error. This parabolic mirror is installed in a fixed position inside a building. Sunlight will be provided by a 32 m² heliostat mirror, equipped with a sun-tracking mechanism, located outside the building (see Fig. 3).

Fig. 3: SOMMER facility

At a direct normal irradiance value of 800 W/m², this arrangement will nominally deliver 10 kW thermal power to a receiver with an area of 100 cm². The heat flux density distribution according to ray-tracing simulations and considering the above receiver area is shown in Fig. 4a. In order to determine the thermal efficiency of a liquid metal-cooled receiver in the focal point, a device for the precise measurement of the incident solar power and its areal distribution on the receiver surface is required, as well as a highly resolved map of local flux densities to characterize the performance of the facility. An assessment of different approaches for measurement solutions resulted in the design of a novel approach for the SOMMER facility: A fast response Heat Flux Micro (HFM) Sensor that is moved across the receiver plane by means of a superimposed linear and a rotational motion component (see Fig. 4b). This design satisfied the demand for short time assembly, testing and commissioning at low costs to obtain precise measurement data.

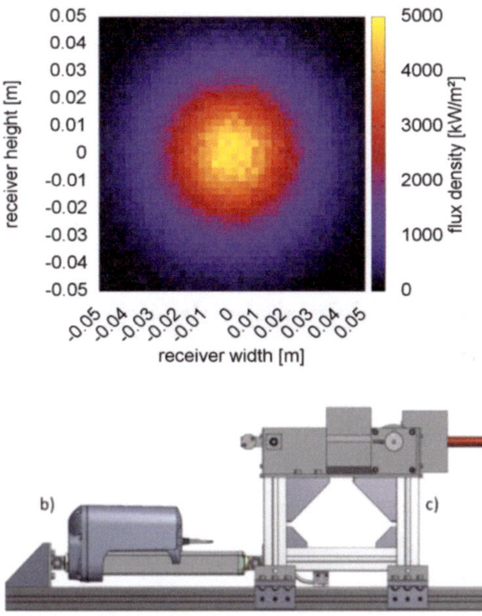

Fig. 4: top) Heat flux density distribution; bottom) Schematic view of the measurement system: a) aluminum support rail b) linear actuator c) car with the rotational gear and the HFM sensor located in the tip of the copper rod

Heavy-liquid metal thermal-hydraulic experiments for safety studies of fast reactors and accelerator-driven systems

Heavy-liquid-metals, such as pure lead and the lead-bismuth eutectic alloy (LBE) are prominent candidate coolants for advanced reactor concepts, which can contribute to the reduction of nuclear waste. In particular, HLMs are considered for the transmutation of long-lived fission products and minor actinides, either in critical reactors or ADSs. In this context, HLM technology is being studied at several institutions worldwide and KALLA has been participating in international research programs since 2002. At a European level, the reference design for this technology is the MYRRHA reactor to be constructed in Belgium.

During 2015 KALLA contributed to the safety analysis of these systems within several European projects with experimental campaigns, including the following highlights.

- A heated 19-rod bundle with wire spacers and cooled by forced-convective LBE was tested in the THEADES LBE loop facility within the FP7 project SEARCH, which finished in April 2015. The measurements showed a high degree of reliability and reproducibility, and valuable lessons for the design and safety analysis of HLM-cooled systems can be derived from the results. The wire spacers, necessary for keeping the short distance between the rods constant, produce a swirl flow that leads to colder regions in the edges and hotter spots in the center. The Nusselt number is in good agreement with the most conservative correlation, see Fig. 5.
- This test section is being modified for further studies within the ongoing FP7 project MAXSIMA. Two small blockage elements covering one sub-channel each are placed at

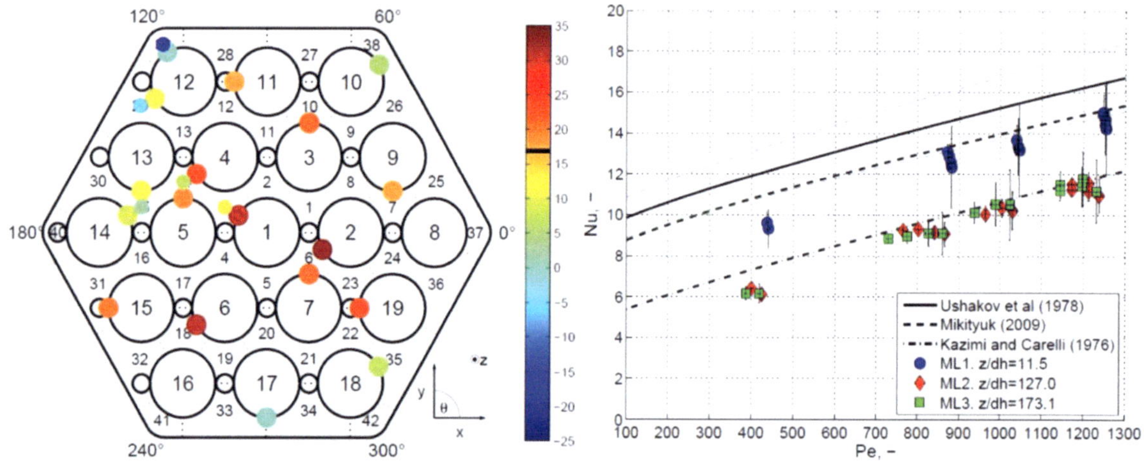

Fig. 5: Experimental results on the heated 19-rod bundle with wire spacers. Left: temperature profiles (above bulk value) for the reference case. Right: measured and predicted Nusselt numbers.

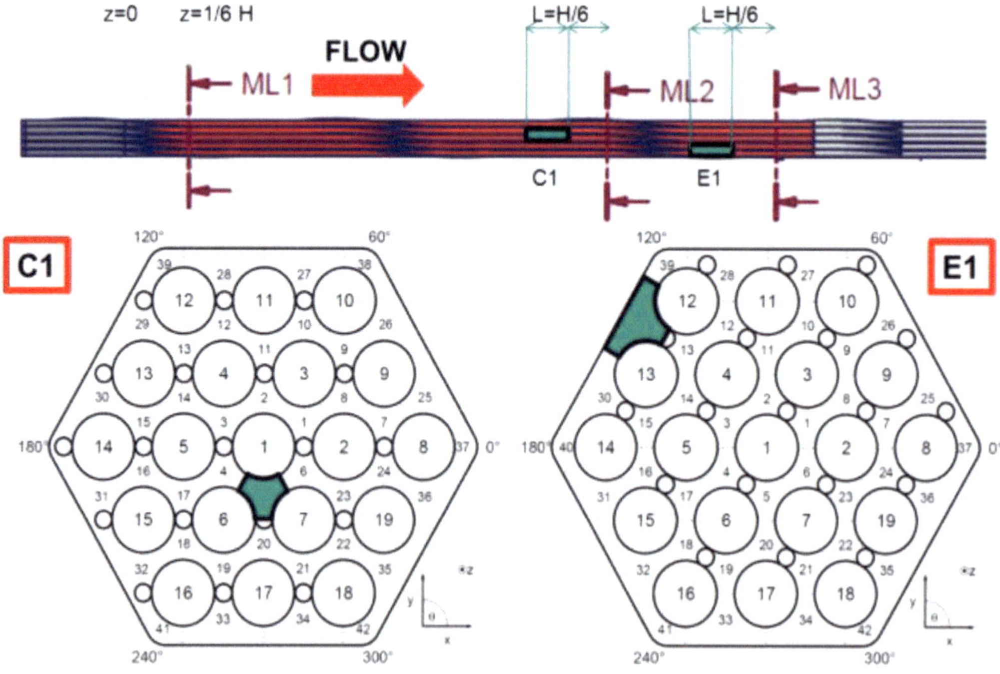

Fig. 6: Location of the blockage elements to be installed in the rod bundle within MAXSIMA (side and top views)

selected locations (one in the center and one in the edge). Additional thermos-couples are installed for detecting the local influence of flow blockages, particularly in respect to hot spots, compared to the reference results obtained within SEARCH, see Fig. 6.

Two new Horizon 2020 projects with participation of KALLA kicked off in April 2015:

- SESAME (2015-2019) is completely dedicated to the thermal-hydraulics studies, considering several liquid-metal reactors designs. At KALLA, the inter-wrapper flow as external mechanism of fuel assemblies during decay heat removal operation shall be tested in a 3-bundle test section, see Fig. 7 top, installed in the THESYS facility. For this campaign, the facility is upgrades with four parallel flow channels and three independent heating sub-systems.

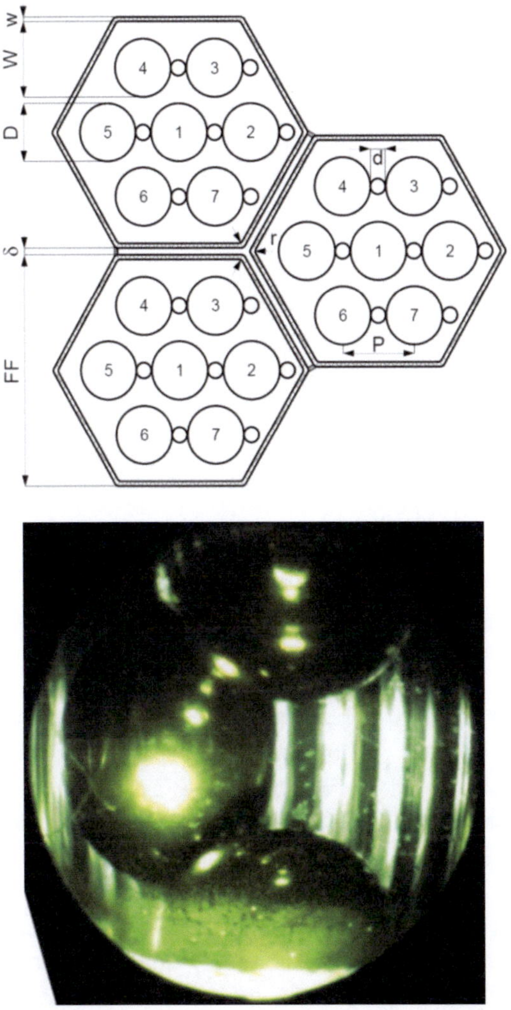

Fig. 7: Top: inter-wrapper flow geometry to be studied within SESAME: gap region between 3 fuel assemblies. Bottom: Porous blockage observed by accumulation of particles in water within MYRTE.

- MYRTE (2015-2019) is focused on the development of MYRRHA, including large work-packages on the accelerator subsystem and core reactor thermal-hydraulics. At KALLA, the process of formation and growth of blockages shall be investigated in water tests, making use of optical instrumentation, see Fig. 7 bottom.

Conclusions

Highlights in the research year 2015 at KALLA have been the principle proof of concept on the feasibility of hydrogen and carbon powder production from methane by thermal cracking in a liquid metal bubble column, very positive evaluation of the LIMTECH results of the first three years as well as granting of two new EU H2020 projects, SESAME and MYRTE.

Again, more than 20 publications and presentations have shown the productivity and the positive spirit of the 15 KALLA team members, including engineers, technicians, PhD students and scientists. One publication with major contributions from KALLA, the "OECD NEA Handbook on Lead-bismuth Eutectic Alloy and Lead Properties, Materials Compatibility, Thermal-hydraulics and Technologies-2015 Edition" has been cited 42 times between its publication in mid-2015 and March 2016, proving the international relevance of liquid metal research and contributions thereof from KALLA.

List of Acronyms

Acronym	Meaning
ADS	Accelerator-driven system
CSP	Concentrated Solar Power
HLM	Heavy liquid metal
KALLA	Karlsruhe Liquid Metal Laboratory
LBE	Lead-bismuth eutectic
MAXSIMA	Methodology, Analysis and experiments for the safety in Myrrha assessment

Acronym	Meaning
MYRRHA	Multi-purpose hybrid research reactor for high-tech applications
SEARCH	Safe Exploitation Related Chemistry for HLM cooled reactors
SOMMER	Solar furnace arrangement with Molten-Metal-cooled Receiver

References

[1] Abánades, A., Rathnam, R.K., Geißler, T., Heinzel, A., Mehravaran, K., Müller, G., Plevan, M., Rubbia, C., Salmieri, D., Stoppel, L., Stückrad, S., Weisenburger, A., Wenninger, H. and Wetzel, T., Development of methane de-carbonisation based on liquid metal technology for CO2-free production of hydrogen. *International Journal of Hydrogen Energy*, pages –, 2015.

[2] Plevan, M.; Geißler, T.; Abanades, A.; Mehravaran, K.; Rathnam, R. K.; Rubbia, C.; Salmieri, D.; Stoppel, L.; Stückrad, S.; Wetzel, T. (2015). Thermal cracking of methane in a liquid metal bubble column reactor: Experiments and kinetic analysis. International Journal of Hydrogen Energy, 40, 8020-8033. doi:10.1016/j.ijhydene.2015.04.062

[3] Geißler, T.; Plevan, M.; Abanades, A.; Heinzel, A.; Mehravaran, K.; Rathnam, R. K.; Rubbia, C.; Salmieri, D.; Stoppel, L.; Stückrad, S.; Weisenburger, A.; Wenninger, H.; Wetzel, T. (2015). Experimental investigation and thermo-chemical modeling of methane pyrolysis in a liquid metal bubble column reactor with a packed bed. International Journal of Hydrogen Energy, 40, 14134-14146. doi:10.1016/j.ijhydene.2015.08.102

[4] Flesch, J.; Fritsch, A.; Cammi, G.; Marocco, L.; Fellmoser, F.; Pacio, J.; Wetzel, T. (2015). Construction of a test facility for demonstration of a liquid lead-bismuth-cooled 10kW thermal receiver in a solar furnace arrangement - SOMMER. Energy Procedia, 69, 1259-1268. doi:10.1016/j.egypro.2015.03.157

[5] Fritsch, A.; Flesch, J.; Geza, V.; Singer, C.; Uhlig, R.; Hoffschmidt, B. (2015). Conceptual study of central receiver systems with liquid metals as efficient Heat transfer fluids. Energy Procedia, 69, 644-653. doi:10.1016/j.egypro.2015.03.074

[6] Marocco, L., Cammi, G., Flesch, J., Wetzel, T., Numerical analysis of a solar tower receiver tube operated with liquid metals, International Journal of Thermal Sciences, Volume 105, July 2016, Pages 22-35

[7] Pacio, J. (Heavy) Liquid metal technology for fast reactors and neutron sources. Landtag Seminar, Goethe Universität Frankfurt am Main, 23 October 2015

[8] Pacio, J., Doolaard, H.; Roelofs, F.; Van Tichelen, K.; Wetzel, Th. Thermal-hydraulic study of the LBE.cooled fuel assembly in the MYRRHA reactor: experiments and simulations. 16th International Topical Meeting on Nuclear Reactor Thermal-hydraulics (NURETH-16), August 30-September 4, 2015, Chicago, USA. Paper number: 12894

[9] Pacio, J.; Litfin, K.; Batta, A.; Viellieber, M.; Class, A.; Doolaard, H.; Roelofs, F.; Manservisi, S.; Menghini, F.; Böttcher, M. (2015). Heat transfer to liquid metals in a hexagonal rod bundle with grid spacers: Experimental and simulation results. Nuclear Engineering and Design, 290, 27-39. doi:10.1016/j.nucengdes.2014.11.001

[10] Pacio, J. C.; Marocco, L.; Stieglitz, R., Low Prandtl-number thermal-hydraulics (2015). Chapter 10 in Handbook of Handbook on Lead-bismuth Eutectic Alloy and Lead Properties, Materials Compatibility, Thermal-hydraulics and Technologies –edited by the OECD-Nuclear Energy Agency.

[11] Pacio, J.; Daubner, M.; Fellmoser, F.; Litfin, K.; Wetzel, T., 8.2 mm rod bundle experimental report, SEARCH Deliverable 2.5., May 2015

[12] Pacio, J.; Daubner, M.; Fellmoser, F.; Litfin, K.; Wetzel, T. Intermediate report on partially blocked wire wrapped LBE rod bundle experiment, MAXSIMA Deliverable 3.3., June 2015

[13] Pacio, J.; Daubner, M.; Fellmoser, F.; Litfin, K.; Wetzel, T. Fuel assembly heat transfer and blockage experiments. In: *SEARCH/ MAXSIMA Technical review meeting, Mol, Belgium, 20-22, April, 2015*

[14] Pacio, J.; Daubner, M.; Fellmoser, F.; Wetzel, T. Inter-wrapper flow experiments. In:

SESAME Technical review meeting, Cadarache, France, 12-14 October 2015

[15] Pacio, J.; Daubner, M.; Fellmoser, F.; Litfin, K.; Wetzel, T. Blockage formation and growth. In: *MYRTE-WP3 Technical review meeting, Cadarache, France, 15 October 2015*

[16] Roelofs, F.; Shams, A.; Pacio, J.; Moreau, V.; Planquart, P.; Van Tichelen, K.; Di Piazza, I.; Tarantino, M. European outlook for LMFR thermal-hydraulics. 16th International Topical Meeting on Nuclear Reactor Thermal-hydraulics (NURETH-16), August 30-September 4, 2015, Chicago, USA. Paper number: 12914

[17] Litfin, K.; Fetzer, J. R.; Batta, A.; Class, A. G.; Wetzel, T. (2015). High power spallation target using a heavy liquid metal free surface flow. Journal of Radioanalytical and Nuclear Chemistry, 305, 795-803. doi:10.1007/s10967-015-4002-z

[18] Bertsche, D.; Knipper, P.; Dietrich, B.; Wetzel, T., The Generalized Lévèque Equation and its application to circular pipe flow, International Journal of Heat and Mass Transfer, 90, 1255-1265, 2015, Pergamon

[19] Bertsche, D.; Knipper, P.; Wetzel, T., "Experimental investigation on heat transfer in laminar, transitional and turbulent circular pipe flow", International Journal of Heat and Mass Transfer, 95, 1008-1018, 2016, Pergamon

[29] Batta, A; Class, AG; Litfin, K; Wetzel, Th; Moreau, V; Massidda, L; Thomas, S; Lakehal, D; Angeli, D; Losi, G; Experimental and numerical investigation of liquid-metal free-surface flows in spallation targets, Nuclear Engineering and Design, 290, 107-118, 2015, North-Holland

[21] Fazio, C.; Sobolev, VP; Aerts, A; Gavrilov, S; Lambrinou, K; Schuurmans, P; Gessi, A; Agostini, P; Ciampichetti, A; Martinelli, L; "Handbook on Lead-bismuth Eutectic Alloy and Lead Properties, Materials Compatibility, Thermal-hydraulics and Technologies-2015 Edition", 2015, Organisation for Economic Co-Operation and Development OECD-NEA

Read more:

[22] Abanades, A.; Geissler, T.; Heinzel, A.; Mehravaran, K.; Müller, G.; Plevan, M.; Rathnam, R. K.; Rubbia, C.; Salmieri, D.; Stoppel, L.; Stückrad, S.; Weisenburger, A.; Wenninger, H.; Wetzel, T. (2015). Development of CO_2-free production of hydrogen by methane decarbonisation based on liquid metal technology. 6th International Conference on Hydrogen Production (ICH2P-2015), Oshawa, CDN, May 3-6, 2015. Proceedings, Paper ID 37.

[23] Geißler, T.; Plevan, M.; Abanades, A.; Mehravaran, K.; Rathnam, R. K.; Rubbia, C.; Salmieri, D.; Stoppel, L.; Stückrad, S.; Wetzel, T. (2015). Experimental analysis of hydrogen production by methane pyrolysis in a liquid metal bubble column. 6th International Conference on Hydrogen Production (ICH2P-2015), Oshawa, CDN, May 3-6, 2015 Proceedings, Paper ID 31.

[24] Marocco, L.; Fellmoser, F.; Flesch, J.; Niedermeier, K.; Wetzel, T. (2015). Liquid metals as efficient heat transfer fluids for next generation CSP systems. Energy, Science and Technology 2015. The energy conference for scientists and researchers. Book of Abstracts, EST, Energy Science Technology, International Conference & Exhibition, 20-22 May 2015, Karlsruhe, Germany, Karlsruhe, KIT.

[25] Pacio, J.; Marocco, L.; Wetzel, T. (2015). Review of data and correlations for turbulent forced convective heat transfer of liquid metals in pipes. Heat and Mass Transfer/Waerme- und Stoffuebertragung, 51 (2), 153-164. doi:10.1007/s00231-014-1392-3

[26 Stoppel, L.; Geißler, T.; Plevan, M.; Wetzel, T.; Abanades, A.; Mehravaran, K.; Rathnam, R. K. (2015). Technical feasibility and design study of methane cracking in a liquid metal. EST 2015 - Energy, Science and Technology, Karlsruhe, May 20-22, 2015 Book of Abstracts.

Decision Support for Disaster Management and Critical Infrastructure Protection

Thomas Münzberg, Wolfgang Raskob

Introduction

Since 2011, the research group Accident Management Systems is carrying out research on decision support tools for disaster management and critical infrastructure (CI) protection. CIs such as the electricity supply, the supply of drinking water, hospitals, pharmacies, dialysis clinics, etc., are organizations and institutions that provide the population with vital services and products. Any disruption of a CI supply can lead to severe consequences, hence, the probability for failures and the potential impacts of CI disruptions have to be mitigated, reduced, or at least managed. Decision support tools for advanced analysis of risk, resilience, and vulnerability are necessary to assist CI providers and disaster management authorities.

In 2015, the undertaken activities by the research group addressed the development of several support tools and analysis methods for different types of CI disruptions and threats. Multiple projects were conducted which dealt with the consequences of power outages (CEDIM activities and Helmholtz Association (HGF) portfolio project "Security Research"), the management of food shortages/ lacks of food (the SEAK project), and the risk and crisis management of terror attacks in public transportation systems (the RIKOV project).

CEDIM Activities: Estimating the Initial Impacts of Power Outages

The Research Group Accident Management Systems is a member of the interdisciplinary research platform of the Center for Disaster Management and Risk Reduction Technology (CEDIM).

As part of the German excellence network of the research program *Integrated Research on Disaster Risk (IRDR)*, CEDIM became an *International Centre of Excellence (ICoE) for Critical Infrastructures and Strategic Planning* in 2015. Embedded in this, two activities were carried out addressing the estimation of initial impacts of power outages and their management at the local level.

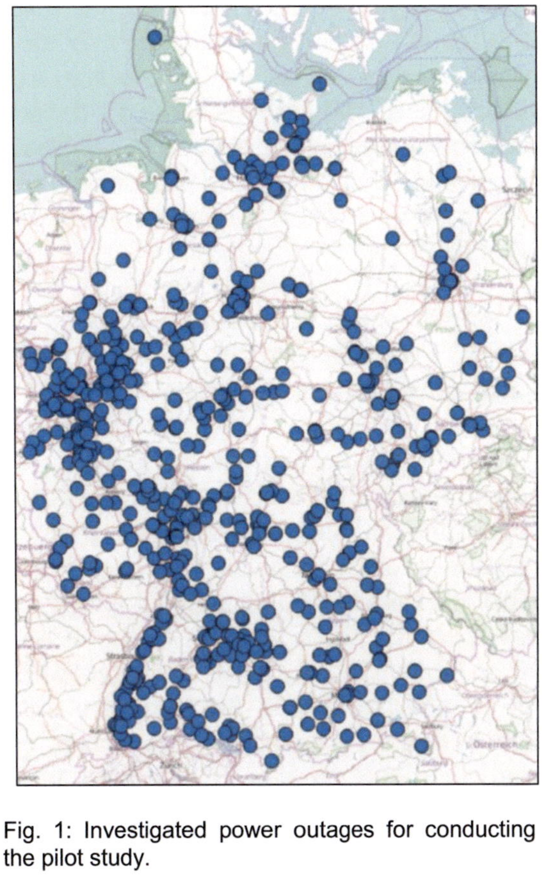

Fig. 1: Investigated power outages for conducting the pilot study.

The first activity aimed at extending the knowledge on power outage impacts by reviewing media articles that reported a power outage incident in Germany. Therefore, a pilot study was conducted that considered how it would be possible to enhance the insights of power outage impacts and how to provide better evidence-based recommendations for disaster management by the evaluation of media data. Derived from a dataset of 234 incidents and associated media articles, information was stored about e.g. the duration, the date and time, the number of affected people, the impacts on CIs, and the consequence-triggered interactions of emergency and disaster management authorities.

The pilot study demonstrated how media articles about power outages were detected [8, 3]. The investigated incidents were reviewed and analyzed. The study showed that it is possible to recognize first tendencies to define the escalation tipping point. Till this escalation level of a power outage, the management of power outage consequences is left to the freedom of the individual. Higher level of escalation may require a response of official authorities. Although the number of samples used in the pilot is too small for final conclusions, the result promised that an enlarged data set may allow disaster management authorities to define, e.g., empirical-based reference scenarios for disaster management planning, protection target levels, and operation criteria for those point in time of a power outage in which further resources are requested.

The second activity in CEDIM addressed an enhanced understanding of concrete power outage consequences in a particular city or county. Therefore, a spatial-temporal vulnerability assessment is developed. The development of a vulnerability assessment aims at estimating the initial sufferings from a power outage in a city or county. This should also enable the building of a common picture of the time-depending effects and a comprehensive view on the functionality of CIs in a municipality. Both is often missing at the local level. The spatial-temporal vulnerability assessment is based on an indicator-based model which takes into account the number, the size, and the type of CIs in the city districts. To consider escalating effects, the model is extended by dynamic parameters that express time-referenced dependencies on electricity and the consumption of coping capacity like emergency power units or batteries which may be implemented at CIs.

In 2015, the research on the vulnerability assessment was focused on the modelling of coping capacity depletion and on macro-micro considerations. The ability to react to potential impacts of power outages depends on the consumption of coping capacity. The more capacity is depleted, the less time is available to plan and realize further response activities and the higher is the vulnerability. The estimation of vulnerability increasing is highly uncertain and depends on the evaluation of the coping capacity consumption. To get better statistical data about this uncertainty, a survey was processed. The survey was embedded in a workshop with representatives from disaster management authorities in the Federal state of Baden-Wurttemberg. The final results were analyzed and published in a book chapter [7]. The results were also used to consider that a CI may consist of different elements which each have an individual preparation level and, hence, the coping capacity is inhomogeneous distributed in a CI [1].

HGF: Modeling Interdependencies of Critical Infrastructures

In 2012, the Helmholtz Association (HGF) initiated the innovative HGF portfolio project "Security Research" in which three Helmholtz Centers, the German Aeronautics and Space Research Centre (DLR), the Research Center Jülich (FZJ) and the Karlsruhe Institute of Technology (KIT) are combining their expert knowledge in cyber security, sensors and platforms, emergency management, and security of CIs in a common research initiative for shaping Critical Infrastructure Protection (CIP). The project aims at establishing a scientific CIP platform which will support CI providers, operators, emergency and disaster management organizations, CIP policy makers, and other scientists whose work is related to CIP.

One of the core activities of the project is the development of an enhanced advising and forecasting capability. This capability will help to investigate the potential occurrence of and the possible prevention strategies against longer lasting power blackouts that may be caused by cyber and terrorist attacks on components of information and communication technology (ICT) and power grids. Therefore, the existing models of each center are methodologically enhanced, and combined taking into account current and upcoming technical trends and systemic developments till 2030. This includes the integration of renewables and the continued decentralized energy generation, improved smart grid capabilities and better stability management through enhanced information and communication technologies. The new technologies imply also new cyber and terrorist threats which may compromise grid stability. Therefore, a fictive cyber-attack on the ICT structure which is operating the smart grid of the city of Karlsruhe is used as a scenario for demonstration. More studies will be carried out to investigate the stability of the energy supply under different conditions, the consequences to CIs and the possible strategies which are applicable to reduce the threat or remediate the consequences.

The task of the Accident Management Group is the development of an agent-based model to consider the potential impacts of a missing power supply on CIs. The agent-based modelling allows to analyze complex systems comprised of individual, autonomous interacting agents. Such agents communicate and negotiate with each other according to their needs and rules, as they try to provide their functionality in a changing environment. In this context, they are used to model particular components of CIs which have autonomy decision-making capability and can react to events triggered by other agents and by that also influence other agent's behaviors. The agent's properties are stored in a relational database and the model framework is coupled to the database which allows storing and reading information related to particular simulations. To characterize the CI's properties, data has to be collected and appropriate rules concerning their behavior in different situations have to be defined. In this way, it is possible to simulate the functionalities of different CIs and their components, their interdependencies, and their behavior in crisis situations. Agent-based modelling has a high level of flexibility and can greatly benefit by combining them with other modeling methods such as system dynamics or decision support methods, e.g. Multi-Criteria Decision Analysis (MCDA).

The core activities in 2015 addressed the fundamental model building including the development of the general CI model frame as well as identifying and collecting the required data for the model implementation [9]. This resulted in a preliminary modelling of CI agents representing some characters of the Karlsruhe water supply and health care facilities. Therefore, an extended literature review for data collection was conducted comprising characteristics and key parameters of the water supply and the health care facilities such as hospitals, mental hospitals, dialysis practices, pharmacies, and other care facilities. 138 relevant facilities were identified and described by individual parameters (e.g. location, size, dependences to other CIs, performance values, etc.). In addition, information and process parameters about some

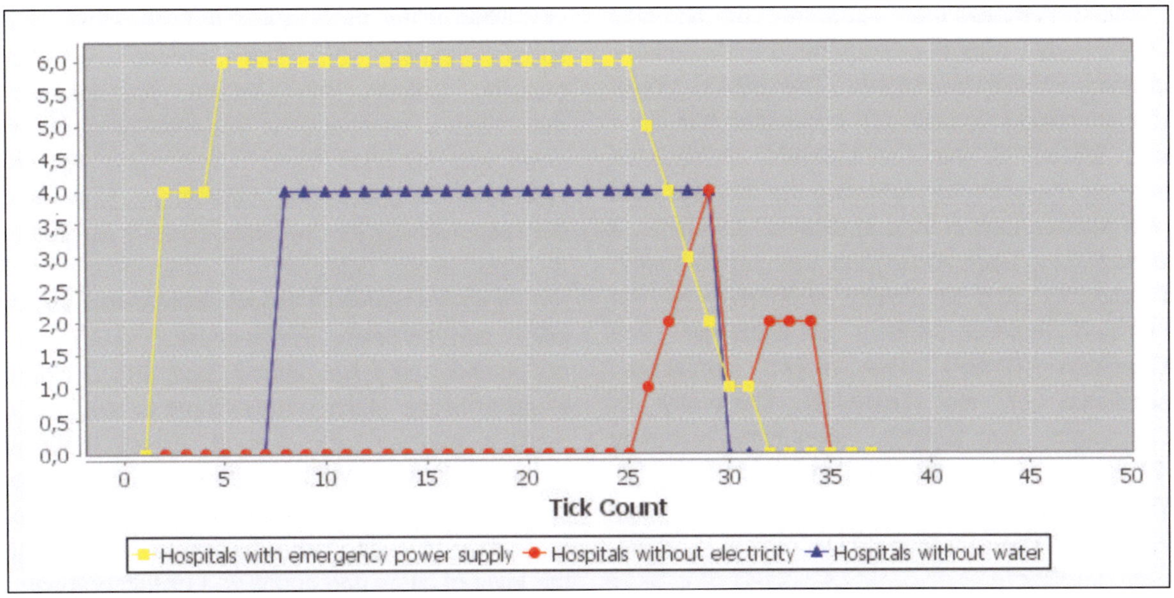

Fig. 2: Visualization of the results of an exemplary power outage and its impacts on other selected CIs.

crisis response management measures related to the above mentioned infrastructures were collected by a literature inquiry.

This allowed a very first demonstration of test simulations which were conducted taking into account some of the identified and described countermeasures [10]. The results allowed conclusions about future research activities. Among others it turned out, that data collection is one of the most crucial steps in setting up the simulation. Therefore, it is important to collect the data as thoroughly as possible and to close gaps in the simplification of the model with carefully defined assumptions that are verified by experts. In a further development step, the CI decision capabilities should be enhanced by using multi-criteria approaches. This would also allow to use additional attributes and to set priorities with changing preferences, in particular from the decision making side of the emergency management teams. Once completed, it will improve the combination of the result of the other research partners in the power grid and the ICT simulations to generate time dependent feedback loops and thus to investigate further attack scenarios and derive possible countermeasures.

SEAK: Decision Support for the Disaster Management of Food Supply Disruption

Food supply chains (FSCs) are optimized systems that are operated on high performance levels. This makes them extremely time and cost effective, but also more vulnerable to disruptions caused by extreme events. The SEAK project (for an introduction see [2]) addressed this issue and aimed at the development of decision support methods for facilitating the management of a sudden lack of food in disaster situations. It is part of the German National Security Research Framework and financially supported by the German Federal Ministry of Education and Research (BMBF). The project consortium consisted of the Technical University of Darmstadt, the logistics consulting company 4flow AG, and the Karlsruhe Institute of Technology (KIT). The project started on January 1, 2013 and was finished on December 31, 2015.

Contributions of the IKET research group Accident Management Systems (UNF) comprised defining and analyzing appropriate threat scenarios as well as compiling a catalog of available measures for authorities for different escalation levels taking into account the respective administrative competences. In the frame of the project, several experts from public authorities have been interviewed by the project team to obtain information related to food supply shortages (e.g. scenario relevant aspects, decision-makers in charge, distribution of responsibilities, possible countermeasures). Based on these findings and on a literature review, three threat scenarios were identified that can imply disruptions of FSCs and seen as highly relevant incidents for the purpose of SEAK. The three scenarios are (i) a heat wave, (ii) a drop out of manpower, and (iii) a failure of the IT system. For each scenario several impact factors were extracted and coupled with the quantitative logistic simulation system that was developed within the project. This allowed the simulation of different aspects considering holistic impacts or specific influences in the provision of food for the population. The results demonstrate how it is possible to identify upcoming shortages in the system and, hence, provide decision support.

Further contributions were the development of two approaches for indicator-based vulnerability assessments. One proposed vulnerability assessment addressed the general topological structure of a food supply system. The results enhanced spatial insights on vulnerabilities by analyzing different regions by their number of inhabitants (demand), the vulnerable groups who are living in the regions and the individual food supply structure [4]. The approach allowed decision makers from disaster management authorities to compare the vulnerability levels of cities and counties in a federal state and to identify static drivers for vulnerability. However, the responsibility for operative disaster response is legislatively delegated to the local crisis management teams at the administrative level of cities and counties. Furthermore, the affected population partly has stored food and drinking water at home which enables them to temporary resist a sudden lack of food. Hence, both a temporal vulnerability evolvement and a higher spatial resolution of vulnerability levels are necessary to support the impact assessment at the level of cities and counties. For this purpose, a second vulnerability assessment was developed [6]. The assessment is based on a

Fig. 3: Representatives from disaster management authorities in Baden-Wurttemberg who participated on the workshop "Disaster Planning for Food Shortages and Power Outages".

review of empirical studies about the behavior of the population in storing food and drinking water. These findings were combined with census data of the population structure of a city or county district and with the results of a workshop with decision makers from disaster management authorities. This information allowed to assess the spatial-temporal vulnerability evolvement based on the population structure and on the consumption of the private food storages.

RIKOV: Evaluating Counter Measures to Cope with the Risks of Terror Attacks on Public Transportation

RIKOV intended to investigate to which extend a holistic risk management approach can enhance protection of public railway transportation systems against terrorist attacks. The project was part of the national research program Research for Civil Security - Security Economics and Security Architecture, initiated and financed by the German Federal Ministry of Education and Research (BMBF). RIKOV started in November 2012 and officially ended in November 2015 with respect to the IKET contributions.

Within the project, the task of IKET was to provide means for prioritization of security measures i.e. to develop methods to provide a ranking of potential security measures in a specific context. The context is on one hand defined by "hard" criteria like the transportation environment of the stakeholder or weapon type used for an potential attack, and on the other hand by "soft" criteria such as acceptance of a security measure by the public (e.g. surveillance cameras). Variations of the criteria are combined in different attack scenarios.

Our basic assumption was to use knowledge from historic events and human experts to provide decision support for selecting appropriate security measures for the different attack scenarios. Because hard and soft criteria had to be combined, we developed a two-step approach using case based reasoning (CBR) as preprocessor for the hard facts and MCDA to adapt the preliminary results to the preferences of the stakeholders/end users.

The first task was to systematically analyze the known attacks on public rail transportation. The data gathered was structured and stored in a knowledge database, including information about potential security measures. A standalone CBR was implemented and adapted to the knowledge database providing a ranking of potential security measures. In the next step the preferences of stakeholders/end users were enquired from experts during a workshop at KIT. The results were stored in a criteria tree structure for a subsequent MCDA. The standalone MCDA tool of IKET was adapted and used to evaluate the stakeholder data.

In 2015 we focused on the final implementation and automatic combination of both methods. As a starting point, a set of scenarios was provided by the RIKOV main application and our applications resulted in a ranked set of security measures which was passed back. We developed and implemented an XML interface for this task. Based on this interface an automated control software was designed, which initiated and conducted the CBR process for each scenario which was defined. The result is taken and combined with additional information from the knowledge database. Both are further used to generate an input file for the prioritization by the MCDA. The result is then passed back as XML to the main RIKOV system.

The approach and its implementation was presented at the Future Security 2015 [3]. The general method was also demonstrated to potential stakeholders/end users at a workshop in Überlingen in October 2015.

References

[1] Brauner, F.; Wiens, M.; Lechleuthner, A.; Münzberg, T.; Fiedrich, F.; Schultmann, F. (2015). Critical infrastructure resilience: a framework for considering micro and macro observation levels. 12th International Conference on Information Systems for Crisis Response and Management (ISCRAM 2015), Kristiansand, N, May 24-27, 2015.Ed.: L. Palen, Palen, L. [Hrsg.] 12th International Conference on Information Systems for Crisis Response and Management (ISCRAM 2015), Kristiansand, N, May 24-27, 2015 Proceedings published online, Online.

[2] Motzke, A., Balster, A., Hansen, O., Herrmannsdörfer, M., Schätter, F., Friedrich, H., Raskob, W., Wiens, M., and Schultmann, F. (2014): The SEAK Project: Decision Support for Managing Disruptions in Food Supply Chains. In: Proceedings of the 9th Future Security, Security Research Conference, Berlin, September 16-18, 2014. Thoma, K.; Häring, I. and Leismann, T. (Eds.), Stuttgart: Fraunhofer Verlag, 2014, pp. 578-581, ISBN 978-3-8396-0778-7.

[3] Müller, T.; Meng, S.; Raskob, W.; Wiens, M.; Schultmann, F. (2015). A transparent approach for prioritising security measures. Security Re-

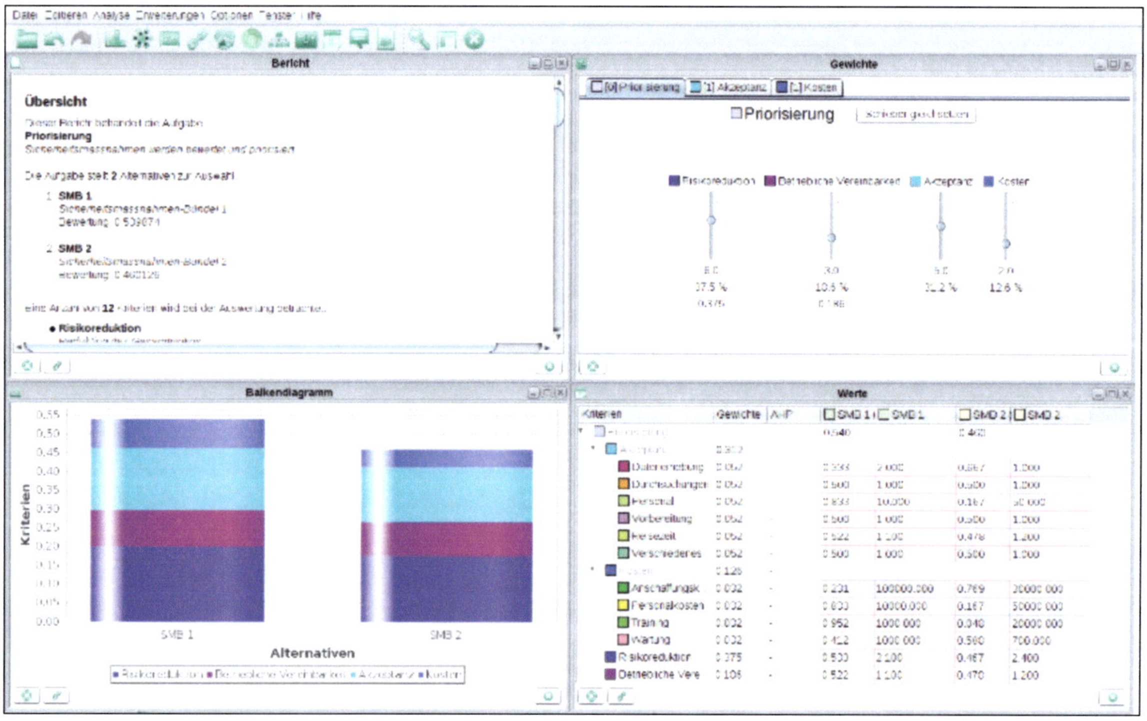

Fig. 4: The Figure displays the four windows of the MCDA user interface during a RIKOV analysis. The right side shows the criteria and their weights. The left side shows the analysis as report and as chart.

search Conference : 10th Future Security, Berlin, September 15-17, 2015. Proceedings. Hrsg.: J. Beyerer, 217-224, Fraunhofer Verl., Stuttgart.

[4] Münzberg, T., Motzke, A., Wiens, M., and Schultmann, F. (2014) Disruptions of Food Supply Chains: A Spatial Vulnerability Assessment Approach for Disaster Management. In: Proceedings of the 9th Future Security, Security Research Conference, Berlin, September 16-18, 2013, Thoma, K., Häring, I., Leismann, T. (Eds.) Stuttgart: Fraunhofer Verlag, pp. 48-55, ISBN: 978-3-8396-0778-7.

[5] Münzberg, T.; Wiens, M.; Raskob, W.; Schultmann, F. (2015). Stromausfall: Schutzziele und Referenzszenarios für kritische Infrastrukturen in der kommunalen Gefahrenabwehrplanung. Gemeinsames Treffen des Arbeitskreises Naturgefahren/-risiken der Deutschen Gesellschaft für Geographie und des Katastrophennetzes KatNET e.V., Berlin, 29.-30. September 2015.

[6] Münzberg, T., Wiens, M., and Schultmann, F. (2015) The Effect of Coping Capacity Depletion on Critical Infrastructure Resilience. In: Proceedings of the 12th International Conference on Information Systems for Crisis Response and Management (ISCRAM-2015), L. Palen, M. Buscher, T. Comes, and A. Hughes, eds., Kristiansand, N, May 24-27, 2015, ISBN: 978-82-7117-788-1.

[7] Münzberg, T., Wiens, M., and Schultmann, F. (2016) Understanding Resilience: A Spatio-temporal Vulnerability Assessment of a Population affected by Sudden Lack of Food, Springer International Series in Operations Research & Management Science: Advances in Managing Humanitarian Operations, 2014, Altay, N., Haselkorn, M., and Zobel, C. (Eds.) ISSN: 0884-8289, Volume 235, pp.257-280, ISSN 0884-8289, DOI 10.1007/978-3-319-24418-1.

[8] Münzberg, T.; Wiens, M.; Raskob, W.; Schultmann, F. (2015). Measuring resilience - the benefits of an empirical study of power outages by media data. Security Research Conference : 10th Future Security, Berlin, September 15-17, 2015 Proceedings. Hrsg.: J. Beyerer, 379-387, Fraunhofer Verl., Stuttgart.

[9] Raskob, W., Bertsch, V., Ruppert, M., Strittmatter, M., Happe, L., Broadnax, B., Wandler, S., and Deines, E. (2015) Security of electricity supply in 2030, Critical Infrastructure Protection and Resilience Europe (CIPRE) Conference & Expo, The Hague, the Netherlands.

[10] Raskob, W.; Wandler, S.; Deines, E. (2015). Agent-based modelling to identify possible measures in case of critical infrastructure disruption. 12th International Conference on Information Systems for Crisis Response and Management (ISCRAM 2015), Kristiansand, N, May 24-27, 2015. Ed.: L. Palen, Online

Read more:

[11] Raskob, W. (2015). How to use decision support systems in a nuclear emergency?. NERIS Workshop 2015, Milano, I, April 27-29, 2015.

[12] Raskob, W.; Schneider, T.; Gering, F.; Charon, S.; Zhelezniak, M.; Andronopoulos, S.; Heriard-Dubreuil, F.; Camps, J. (2015). PREPARE: innovative integrated tools and platforms for radiological emergency preparedness and post-accident response in Europe. Radiation Protection Dosimetry, 164, 170-174. doi:10.1093/rpd/ncu279

Group: Hydrogen

Hydrogen Risk Assessment for Nuclear Applications and Safety of Hydrogen as an Energy Carrier

Andrei Denkevits, Benedikt Hoess, Andrea Hurtado, Olaf Jedicke, Thomas Jordan, Birgit Kaup, Ke Ren, Alexei Kotchourko, Mike Kuznetsov, Alexander Lelyakin, Frank Prestel, Michael Ranft, Reinhard Redlinger, Maria Stassen, Jianjun Xiao, Zhanjie Xu, Jorge Yanez, Yu, Zhang,

Introduction

In 2015 the Hydrogen Group continued to develop the in-house CFD HyCodes, GASFLOW and COM3D (see [1] to [6]), and conducted several experimental programs to develop further the criteria for flame acceleration and deflagration-detonation-transition or to provide model vali-dations data. Still the code development is accommodated in the nuclear program NUSAFE and therefore suffers from continuous budget constraints. However, the quite successful licen-sing of the HyCodes to external users provides additional financial resources for the development work.

The next three chapters describe three selected projects in more detail. First, a dedicated fluid-structure-coupling model development for COM3d was supported directly by automotive in-dustry. Therefore, there has been no related publication so far. The next example addresses experiments for inhomogeneous flat layers of premixed gases co-funded by GRS. Although these large scale experiments provide obvious links to the Fukushima accidents, the results ser-ve in a more general way as reference data and for model development. The last example refers to experiments of unconfined spherical pre-mixed clouds prepared in soap bubbles.

The selected references at the end of this section are directly referring to the two latter projects.

Coupling Flame and Structural Dynamics

In the frames of general COM3D code improvement and supplemented model development during last year, a concept of vo-lume porosity together with model of moving walls were elaborated and implemented into the code.

These new attainments allowed to realize support of real-time data exchange with finite-element analysis code ABAQUS - © Dassault Systèmes providing possibility to perform simulations of the gas-dynamic process-ses simultaneously with possible adaptation of the geometrical and thermo-dynamical environ-mental conditions.

The current implementation of the COM3D and ABAQUS two-way coupling provides exchange of information between these two codes on every time step and on the pre-defined set of nodes known for both codes. From COM3D, the information on the 3D force acting on each node is transferred, and in the opposite direction from ABAQUS, the 3D displacement vector and 3D velocity vector are transferred.

For the accounting of moving walls the concept of 'moving embedded boundaries' was realized in COM3D. In the frames of this concept, it is assumed that the calculating mesh is re-generated only when displacement exceeds the size of a cell, otherwise accounting of the wall dislocation is performed using sub-grid model of the 'embedded' boundary. 'Embedded' boundary model includes control of the calculation cell fraction occupied by gas/solid material, and corresponding correction of fluxes through the cell boundaries and field parameters inside the cell.

Currently the model takes into account only the one-dimensional movements, which are directed along the normal to the moving boundary, and therefore only components normal to the surface of exchanged fields are participated in the data transfer between codes.

For the class of problems, which are of interest in the co-simulation projects, relatively rapid changes of the given model geometry are assumed. Therefore, as a partner for COM3D common simulation agent, the ABAQUS/Explicit code variant was selected.

In the data exchange procedure, three individual programs are taking part: ABAQUS/Explicit, ABAQUS CSE Co-Simulation Director and COM3D. Therefore, newly developed COM3D code extension can be used when ABAQUS license is available, since first two packages are included into standard ABAQUS distribution suite.

For successful initiation and performance of COM3D-ABAQUS complex, it is necessary to create consistent geometry models, which will be accepted by both codes, and in both codes an interface surface(s), on which data exchange can be performed, should be defined in terms specific for each participant.

Fig. 1a: Experimental set-up for investigating transient flame behavior coupled to flexible walls using transparent walls and optical measurement techniques in HYKA Test Chamber Q160

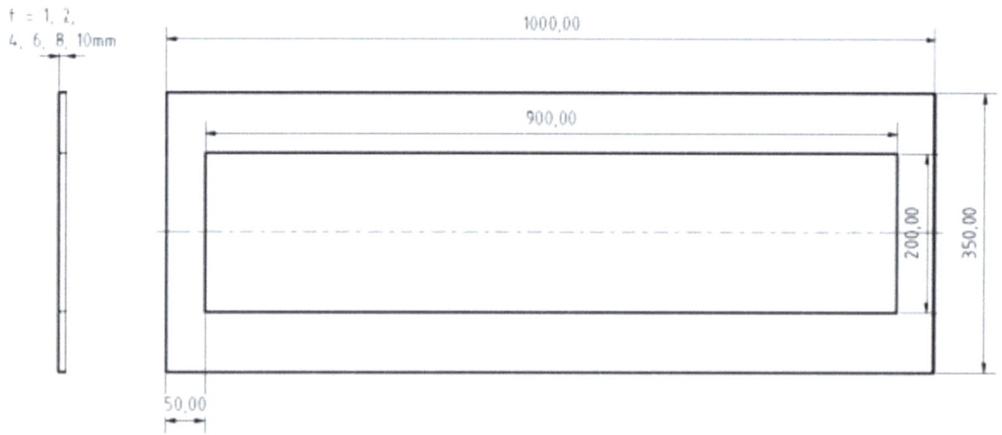

Fig. 1b: Schematic of the experimental transparent combustion channel.

The COM3D conventional geometry model (i.e., without peculiar for co-simulation specifications, such as data exchange surface(s)) can be created using conventional ways, e.g., by definition in terms of geometry data language (GDL), by import from STL data (ABAQUS code has possibility to export STL), and using one fully new specially developed for communication with ABAQUS method. The last one is based on the import of ABAQUS model data, extracted from ABAQUS model database by specialized COM3D utility. This utility provides export of exchange surface geometry data as well.

The full functionality of COM3D-ABAQUS complex is supported including restart possibility, online execution control and visualization. User's manual is available.

Results

The new developments were validated against experiments performed in the Hydrogen Group. The experimental set-up is shown in Figure 1a.

Here only one test case will be described, namely Test 40. This test was accelerated combustion with deflagration to quasi-detonation regime. The combustion was initiated in flat channel (see Fig. 1b) of 10 mm height and with metal grid inside volume for promoting flame acceleration. The channel is confined by Plexi-glas walls of 8 mm thickness.

The combustion was modelled using standard COM3D flame acceleration model KYLCOM+. Initial combustion was slow and the displacement of walls was affected by uniform pressure distribution.

In Figure 2 and Figure 3 one can see the effect of the pressure on wall displacement.

Comparison of the data from COM3D (Fig. 3) and from ABAQUS (Fig. 4) confirm exact co-simulation program complex operation. Utilization of the developed technique allowed obtaining results unreachable without accounting on wall displacements. Comparison of the experimental results and data from simulation with and without displacement demonstrated massive over-estimation of the pressures observed during flame propagation.

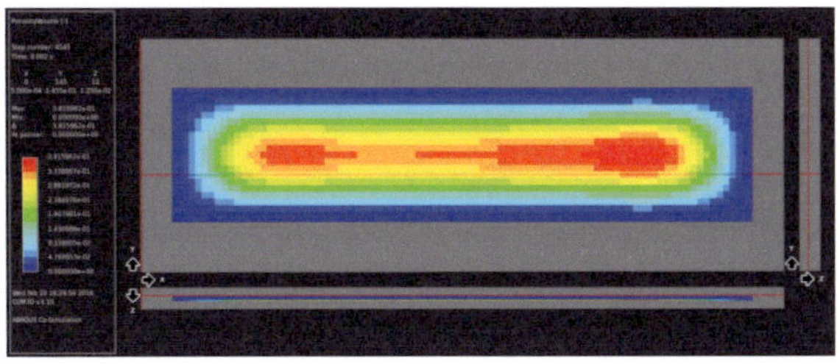

Fig. 2: Wall displacement distribution by uniform overpressure.

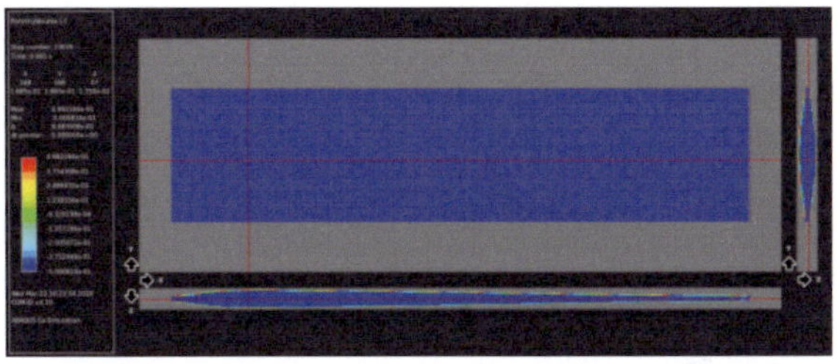

Fig. 3: Wall displacement at the end of strong flame acceleration and quasi-detonation in the second half of the facility in COM3D. Time 5 ms after beginning of the process.

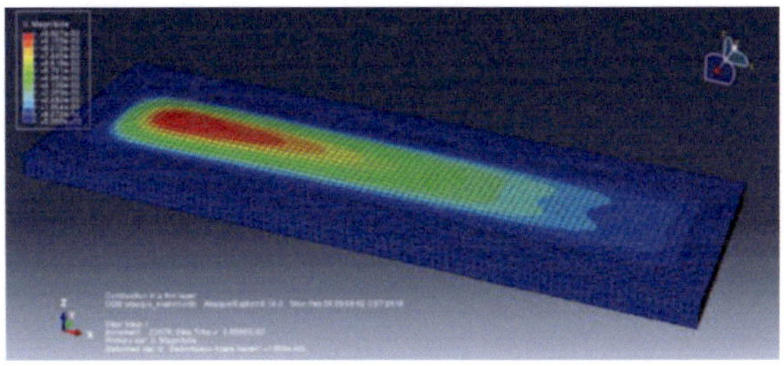

Fig. 4: Wall displacement at the end of strong flame acceleration and quasi-detonation in the second half of the facility in ABAQUS. Time 5 ms after beginning of the process.

The pressure prediction obtained in the case with wall displacement accounting demonstrated excellent agreement (see Figure 5) between experiment and simulations.

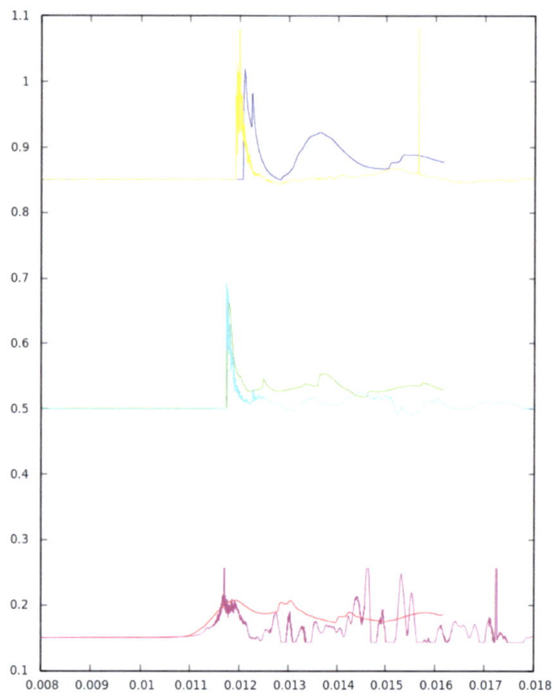

Fig. 5: Comparison of the experimental results and simulations as X-t diagram. Records of the three pressure transducers are shown. The horizontal axe corresponds to time in s, vertical denotes position of the transducer

Acknowledgements
The work was supported by Toyota Tsusho Europe S.A.

Transient Flame Behaviour in Partially Confined Flat Gas Layers

In the frame of the BMWi financed project ("GRS Vorhaben 1501426") the Hydrogen Group has continued experimental work related to the flame behavior in partially confined layers of premixed gases with a concentration gradient.

The experiments conducted in the horizontal configuration in the first phase of the program before 2015 suggested the "concept of maximum concentration". It claims that homogenous mixtures and gradient mixtures produce the same flame velocities and overpressure effects if the maximum – and not the average – concentration in the gradient mixture equals the concentration of the homogenous mixture.

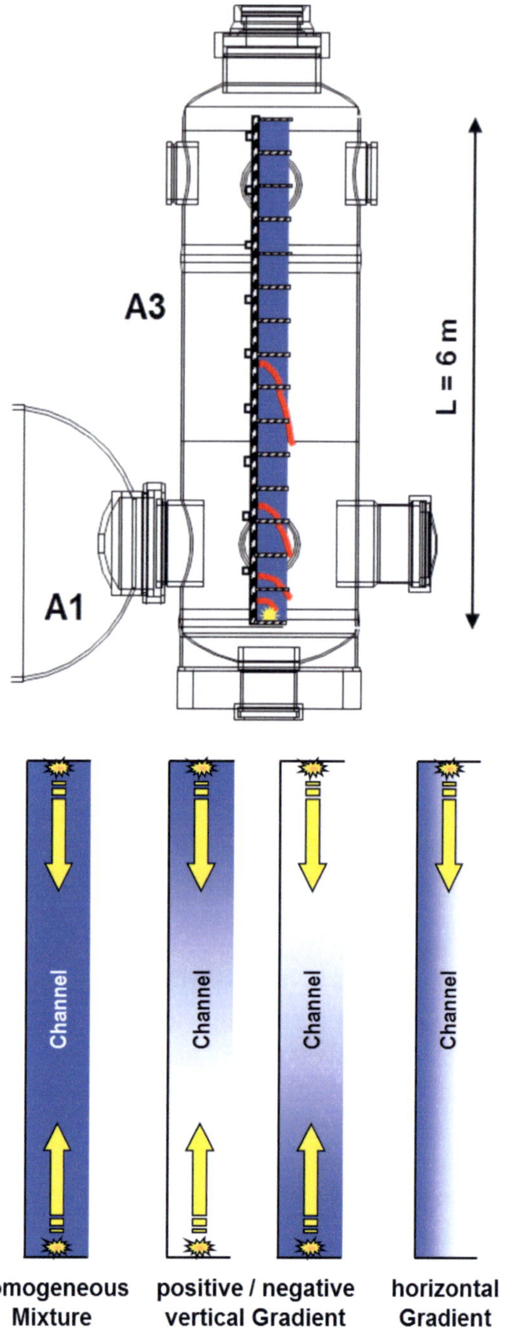

Fig. 6: Arrangement of the semi-open test channel in HYKA V30 and principle variations in the experimental matrix

The focus in 2015 was on layers attached to vertical walls and to check whether the maximum concentration concept holds for these configurations. The experiments were conducted in the HYKA test vessel V30. The experimental matrix included homogeneous mixtures and pre-mixed layers with vertical and horizontal concentration gradients. Besides hydrogen inventory, ignition point and blockage ratio in the driver section have been varied.

Results

The results of all tests with vertical arrangement are shown in two charts in the figure below. In the upper chart the maximum pressure generated by the combustion is shown depending on the maximum concentration in the premixed layers before ignition, suggesting that the above concept of maximum pressure holds. The light and dark blue curves represent the experiments with homogeneous mixtures. The lower chart presents the identical results in the more intuitive way: the measured maximum pressure is now plotted over the averaged concentration. It is easy to see that the latter way of presentation gives much less scatter compared to applying the maximum concentration concept.

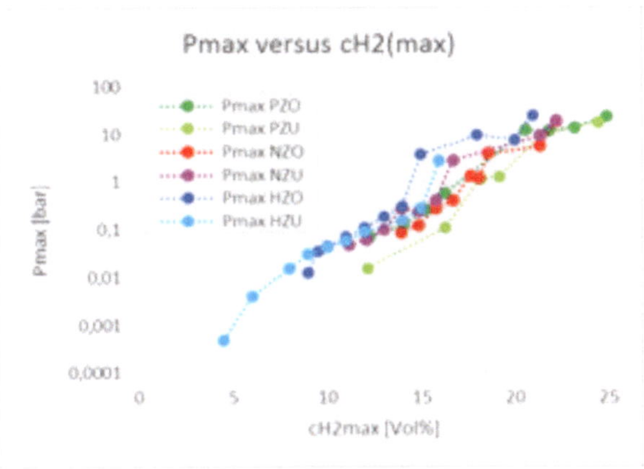

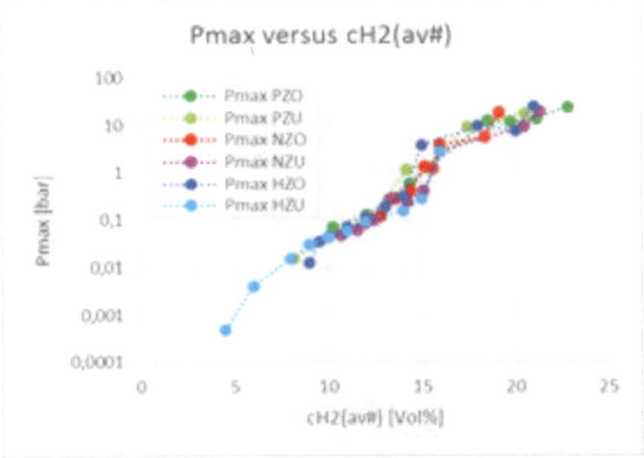

Fig. 7: Maximum pressure of the vertical channel combustion tests over maximum concentration (top) and over average concentration (bottom)

The convergence for the average concentration chart clearly indicates, that the maximum concentration criteria, derived from previous experiments, is less suitable compared to the standard criteria, based on the average concentration, at least for these vertically arranged layers.

It is planned to extend these kind of investigations including carbon monoxide in the premixed systems. Carbon monoxide points to safety issues related to the late phase of severe accidents, including Fukushima, and related to synthesis gas in the chemical industry.

Acknowledgements
The high quality experiments were conducted by Pro-Science and the work supported by the german BMWi under the title “GRS Vorhaben 1501426”.

Experimental study of flame instabilities in unconfined geometries

Hydrogen risk assessments of prototypical scenarios associated with a wide use of hydrogen as an energy carrier typically assume leakage of the gas, premixing with air and ignition to form a more or less unconfined 3d combustion process. Many studies have already analyzed the behavior of spherical flames, however, in most of them a truly unconfined environment could not be simulated, given that a plastic cover was used to contain the gas, which affects the development of the flame.

Therefore, the purpose of this specific study was to design an experimental setup, which approaches much better fully unconfined conditions, and to analyze the effects of instabilities on the flame dynamics. The soap bubble method was chosen, as it is the technique that most accurately simulates an unconfined geometry.

The experimental facility was constructed and in total 55 experiments were conducted, in which the soap bubbles were filled with different mixtures, namely hydrogen-air, hydrogen-oxygen, methane-air and methane-oxygen, in different concentrations. Shadowgraph method was used to capture the whole process in high-speed videos with frame rates up to 30000 fps.

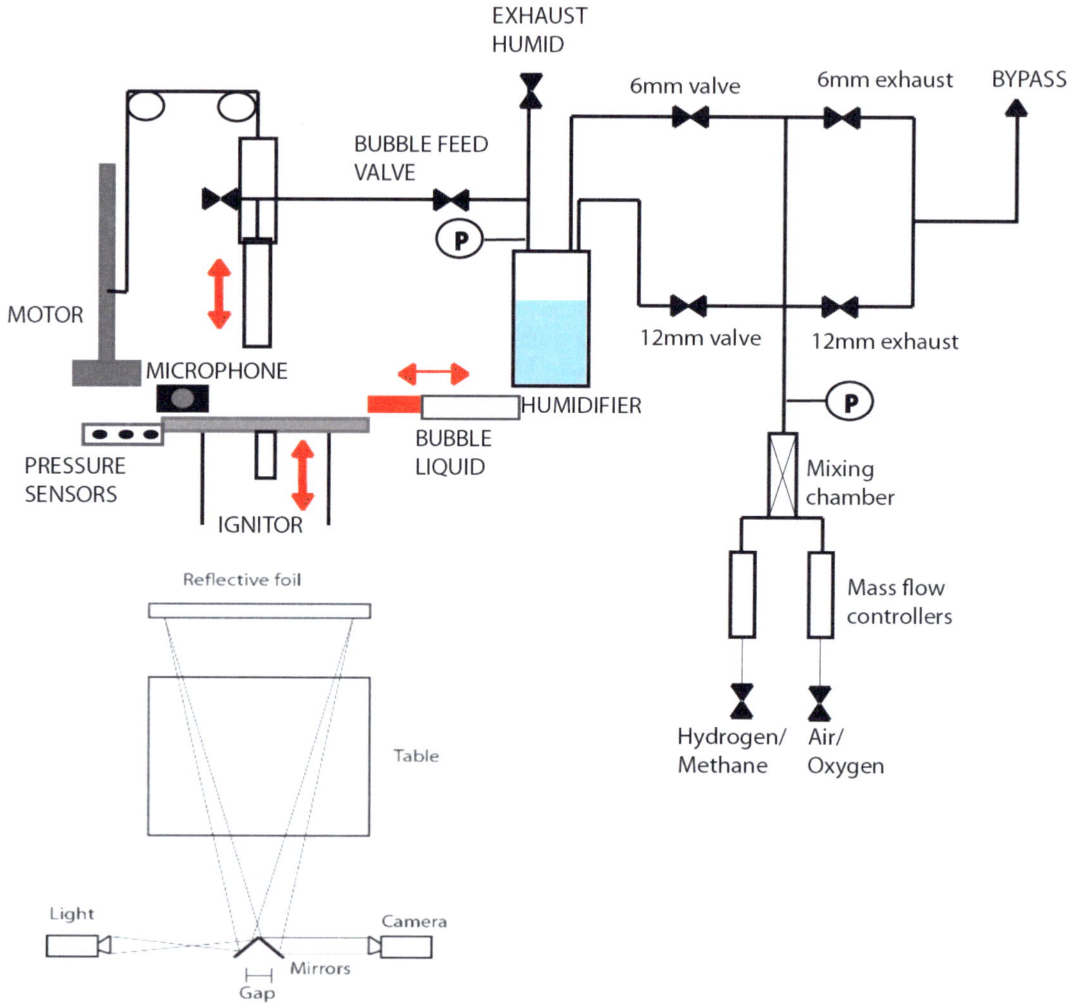

Fig. 8: Experimental set-up for the unconfined soap bubble experiments including the optical measurement arrangement (top view)

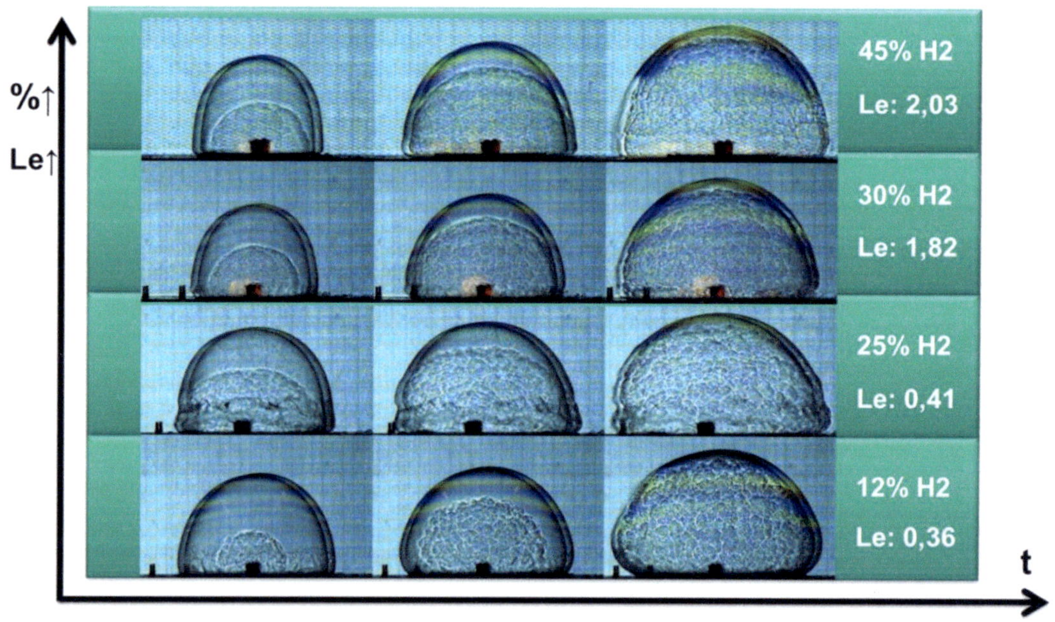

Fig. 9: Influence of Lewis number Le on flame speed (flame indicated by the light semi spheres within the bold premixed bubble) and on flame instabilities (indicated by the wrinkling of the flame surface)

The relationship between Lewis, Markstein and Peclet numbers and flame instabilities, specifically wrinkling, as well as flame speed, were analyzed through the examination of the obtained high-speed movies. Overpressure outside the premixed bubble was measured and compared to the results of the TNO Multi-Energy Model. Parameters for Deflagration-to-Detonation Transition were studied, especially in those experiments in which obstacles were added in the center to promote flame acceleration.

Results

The main findings of this experimental program have been that the measured flame speed fits well the expected behavior and the flame speed rises considerably by adding obstacle in the premixed bubbles. Only deflagrations have been observed in the unconfined configurations. The flame surface wrinkling can be well predicted with the help of the Lewis number Le. The unstable nature of $H_2/O_2/N_2$ combustion leads to an incompatibility with the TNO Model. The TNO model is based on hydrocarbon fuels and fuel gases and therefore does not fit.

References

[1] Xiao, J., Travis, J. R., Royl, P., Kuznetsov, M., Redlinger, R., Svishchev, A., Necker, G., Jordan T., Breitung, W., GASFLOW-MPI: a Parallel Pressure-based Semi-implicit All-speed CFD Code for Hydrogen Explosion Risk Analysis, Accepted for publishing, Nuclear Engineering and Design, 2015

[2] Xiao, J., Travis, J. R., Redlinger, R., Jordan, T., Kuznetsov, M., Svishchev, A., Breitung, W., (2015), Three dimensional scalable all-speed CFD code GASFLOW-MPI: Applications to turbulent combustion of premixed hydrogen-air mixtures with heat and mass transfer. 16th International Topical Meeting in Nuclear Reactor Thermal Hydraulics (NURETH-16), Chicago, Ill, August 30 - September 4, 2015 Proceedings, 4516-4528, ANS, LaGrange Park, Ill.

[3] Xiao, J., Travis, J. R., Royl, P., Svishchev, A., Jordan, T., Breitung, W. (2015), PETSC-based parallel semi-implicit CFD code GASFLOW-MPI in Application of hydrogen safety analysis in containment of nuclear power plant. 1st Joint International Conference on Mathematics and Computations (M&C), Supercomputing in Nuclear Applications (SNA) and Monte Carlo (MC) Method, Nashville, Tenn., April 19-23, 2015 Proceedings, 1279-1292, ANS, LaGrange Park, Ill.

[4] Jordan, T., Royl, P., Kotchourko, A., Lelyakin, A., Kim, B. J., Moon, Y. T., Hwang, D. H. (2015), Use of the KIT HYCODES system with a conservative source for the simulation of hydrogen distribution and combustion with/ without spray in a large break LOCA in the Korean APR1400. 46. Jahrestagung Kerntechnik, Berlin, 5.-7.Mai 2015, CD-ROM, INFORUM GmbH, Berlin.

[5] Xiao, J., Travis, J. R., Royl, P., Necker, G., Svishchev A. and Jordan T., "GASFLOW-MPI: A Scalable Computational Fluid Dynamics Code, VOLUME 1: Theory and Computational Model", Revision 1.0, KIT Scientific Report Nr. 7710, KIT Scientific Publishing, Karlsruhe, October 2015.

[6] Xiao, J., Travis, J. R., Royl, P., Necker, G., Svishchev, A. and Jordan, T., "GASFLOW-MPI: A Scalable Computational Fluid Dynamics Code, VOLUME 2: Users' Manual", Revision 1.0, KIT Scientific Report Nr. 7711, KIT Scientific Publishing, Karlsruhe, October 2015.

[7] Kuznetsov, M., Friedrich, A., Stern, G., Kotchourko, N., Jallais, S., L'Hostis, B. (2015), Medium-scale experiments on vented hydrogen deflagration. Journal of Loss Prevention in the Process Industries, 36, 416-428. doi:10.1016/j.jlp.2015.04.013

[8] Kljenak, I., Kuznetsov, M., Kostka, P., Kubisova, L., Maltsev, M., Manzini, G., Povilaitis, M. (2015), Simulation of hydrogen deflagration experiment - Benchmark exercise with lumped-parameter codes. Nuclear Engineering and Design, 283, 51-59. doi:10.1016/j.nucengdes.2014.07.016

[9] Yanez, J., Kuznetsov, M., Grune, J. (2015), Flame instability of lean hydrogen-air mixtures in a smooth open-ended vertical channel. Com-

bustion and Flame, 162, 2830-2839. doi:10.1016/j.combustflame.2015.04.004

[10] Xiao, J., Travis, J. R., Kuznetsov, M. (2015), Numerical investigations of heat losses to confinement structures from hydrogen-air turbulent flames in ENACCEF facility. International Journal of Hydrogen Energy, 40, 13106-13120. doi:10.1016/j.ijhydene.2015.07.090

[11] Yanez, J., Kuznetsov, M., Souto-Iglesias, A. (2015), An analysis of the hydrogen explosion in the Fukushima-Daiichi accident. International Journal of Hydrogen Energy, 40, 8261-8280. doi:10.1016/j.ijhydene.2015.03.154

[12] Grune, J., Breitung, W., Kuznetsov, M., Yanez, J., Jang, W., Shim, W. (2015), Flammability limits and burning characteristics of CO-H_2O-CO_2-N_2 mixtures at elevated temperatures. International Journal of Hydrogen Energy, 40, 9838-9846. doi:10.1016/j.ijhydene.2015.06.004

[13] Xu, Z.,; Jordan, T. (2015), GASFLOW Simulations for Containment Venting System of Emsland Nuclear Power Plant (KIT Scientific Reports; 7695). KIT Scientific Publishing, Karlsruhe. doi:10.5445/KSP/1000046875

[14] Xu, Z., Jordan, T. (2015), Hydrogen risk analysis for a generic nuclear containment ventilation system. Proceedings of the 6th International Conference on Hydrogen Safety (ICHS 2015), Yokohama, J, October 19-21, 2015. Ed.: N. N. Carcassi, CD-ROM.

[15] Benteboula, S., Malet, J., Bleyer, A., Bentaib, A., Paladino, D., Guentay, S., Andreani, M., Tkatschenko, I., Brinster, J., Dabbene, F., Kelm, S., Allelein, H. J., Visser, D. C., Benz, S., Jordan, T., Liang, Z., Kiselev, A., Yudina, T., Filippov, A.; Khizbullin, A.; Kamnev, M.; Zaytsev, A., Loukianov, A. (2015), EU-ERCOSAM project. Scaling from nuclear power plant to experiments. International Congress on Advances in Nuclear Power Plants (ICAPP 2015), Nice, F, May 3-6, 2015 Proceedings on USB-Stick, Paper 15378, Madison, Wis.: Omnipress, 2015.

[16] Malet J., Bleyer A., Bentaib A., Paladino D., Guentay S., Andreani M., Tkatschenko I., Brister J., Dabbenne F., Kelm S., Gehr R., Allelein H.-J., Visser D.C., Benz S., Jordan T., Liang Z., Kiselev A., Yudina T., Filippov A., Khizbullin A., Kamnev M., Popova T., Zaytsev A., HYMIX benchmarking tests: code-to-code comparison, Proceedings of the 2015 International Congress on the Advances in Nuclear Power Plants, Nice, France 3-6 May 2015

[17] Kuznetsov, M., Yanez, J., Grune, J., Jordan, T., Friedrich, A. (2015), Hydrogen combustion in a flat semi-confined layer with respect to the Fukushima Daiichi accident. Nuclear Engineering and Design, 286, 36-48. doi:10.1016/j.nucengdes.2015.01.016

[18] Jordan, T. (2015), Hydrogen safety. Lessons learnt from nuclear safety. 3rd Hydrogen and Fuel Cells Conference, Cancun, MEX, November 17-20, 2015.

[19] Grune, J., Sempert, K., Friedrich, A., Kuznetsov, M., Jordan, T. (2015), Detonation wave propagation in semi-confined layers of hydrogen-air and hydrogen-oxygen mixtures. Proceedings of the 6th International Conference on Hydrogen Safety (ICHS 2015), Yokohama, J, October 19-21, 2015. Ed.: N. N. Carcassi, CD-ROM.

[20] Yanez, J., Kuznetsov, M., Lelyakin, A., Experimental study and theoretical analysis of a ‘strange wave’ behavior, Proceedings of the European Combustion Symposium, Budapest, Hungary, 30 March–2 April 2015, paper P4-11.

[21] Yanez, J., Kuznetsov, M., Lelyakin, A., An Effect of Detailed Chemistry on Flame Acceleration and Detonation Preconditioning, Proceedings of the European Combustion Symposium, Budapest, Hungary, 30 March–2 April 2015, paper P4-12.

[22] Kuznetsov, M., Grune, J., Kudriakov, S., Studer, E. (2015), Experimental investigation of H_2-air deflagration in the presence of longitudinal concentration gradients. International Congress on Advances in Nuclear Power Plants (ICAPP 2015), Nice, F, May 3-6, 2015 Proceedings on USB-Stick, Paper 15360, Madison, Wis.: Omnipress, 2015.

[23] Kuznetsov, M., Lelyakin, A., Alekseev, V., Matsukov, I., Detonation Transition in Relatively Short Tubes, Proc. of the 30th International Symposium on Shock Waves (ISSW30), July 19-24, 2015, Tel-Aviv, Israel. Paper #600, p. 2

[24] Yanez, J., Kuznetsov, M. (2015), An analysis of flame instabilities based on Sivashinky equation. Radulescu, M.I. [Hrsg.] 25th International Colloquium on the Dynamics of Explosions and Reactive Systems (ICDERS 2015), Leeds, GB, August 2-7, 2015 Proceedings publ.online Paper No.261.

[25] Makarov, D., Kuznetsov, M., Molkov. V., Vented Deflagrations of Localized Uniform Hydrogen-Air Mixtures. Ninth Mediterranean Combustion Symposium, 7–11 June 2015, Rhodes, Greece, FR-9, 12 p.

[26] L'Hostis, B., Houssin-Agbomson, D., Jallais, S., Vyazmina, E., Dang-Nhu, G., Bernard-Michel, G., Kuznetsov, M., Molkov, V., Chernyavskiy, B., Shentsov, V., Makarov, D., Dey, R., Hooker, P., Baraldi, D.,; Weidner, E., Melideo, D., Palmisano, V., Venetsanos, A., Der Kinderen, J. (2015), Guidelines and recommendations for indoor use of fuel cells and hydrogen systems. Carcassi, N.N. [Hrsg.] Proceedings of the 6th International Conference on Hydrogen Safety (ICHS 2015), Yokohama, J, October 19-21, 2015 CD-ROM Paper ID 209.

[27] Kuznetsov, M., Grune, J., Tengah, S., Yanez, J. (2015), Exerimental study of 2D-instabilities of hydrogen flames in flat layers. Radulescu, M.I. [Hrsg.] 25th International Colloquium on the Dynamics of Explosions and Reactive Systems (ICDERS 2015), Leeds, GB, August 2-7, 2015, Proceedings publ. online Paper No.313.

[28] Jordan, T. (2015), Sicherheitsaspekte bei der Erzeugung, Speicherung, Verteilung und Nutzung von Wasserstoff. 11.Treffen der Expertengruppe H2-System, Energieagentur NRW, Düsseldorf, 19.Februar 2015.

[29] Jordan, T., Head of the Hydrogen Safety Working Group at the Karlsruhe Institute of Technology (KIT), Dr Thomas Jordan talks to SETIS, https://setis.ec.europa.eu/publications/setis-magazine/fuel-cells-and-hydrogen/head-of-hydrogen-safety-working-group-karlsruhe

Read more:

[30] Denkevits, A.; Hoess, B. (2015), Hybrid H_2/Al dust explosions in Siwek sphere. Journal of Loss Prevention in the Process Industries, 36, 509-521. doi:10.1016/j.jlp.2015.03.024

[31] Jordan, T. (2015,. KIT fueling station experience. 6th International Conference on Hydrogen Safety (ICHS 2015), Yokohama, J, October 19-21, 2015.

[32] Kuznetsov, M., Grune, J. (2015), Flame acceleration and deflagration-to-detonation transition in a torus geometry. 12th Inter. Conference on Flow Dynamics (ICFD 2015), Sendai, J, October 27-29, 2015 Proceedings on USB-Stick, 224-225.

[33] Kuznetsov, M., Pariset, S., Friedrich, A., Stern, G., Travis, J., Jordan, T. (2015) Experimental investigation of non-ideality and non-adiabatic effects under high pressure re-leases. International Journal of Hydrogen Energy, 40, 16398-16407 doi:10.1016/j.ijhydene.2015.10.037

[34] Kuznetsov, M.; Yanez, J. (2015). An analysis of the hydrogen explosion in the Fukushima-Daiichi accident. 12th International Conference on Flow Dynamics (ICFD 2015), Sendai, J, October 27-29, 2015 Proceedings on USB-Stick, 296-297.

[35] Kuznetsov, M., Yanez, J. (2015), Flame instability in a mm-scale layer. 12th International Conference on Flow Dynamics (ICFD 2015), Sendai, J, October 27-29, 2015 Proceedings on USB-Stick, 230-231.

[36] Makarov, D., Molkov, V., Hooker, P., Kuznetsov, M. (2015), Venting deflagrations of local hydrogen-air mixture. Carcassi, N.N. [Hrsg.] Proceedings of the 6th International Conference on Hydrogen Safety (ICHS 2015), Yokohama, J, October 19-21, 2015 CD-ROM Paper ID 196.

[37] Malet, J., Mimouni, S., Manzini, G., Xiao, J., Vyskocil, L., Siccama, N.B., Huhtanen, R. (2015), Gas entrainment by one single French PWR spray, SARNET-2 spray benchmark. Nuclear Engineering and Design, 282, 44-53. doi:10.1016/j.nucengdes.2014.12.008

[38] Redlinger, R. (2015), Numerical simulation of hybrid dust/gas explosion experiments in the standard 20-L sphere. Fusion Engineering and Design, 100, 419-424. doi:10.1016/j.fusengdes.2015.07.006

[39] Shentsov, V., Kuznetsov, M., Molkov, V. (2015), The pressure peaking phenomenon: Vali-

dation for unignited releases in laboratory-scale enclosure. Carcassi, N.N. [Hrsg.] Proceedings of the 6th International Conference on Hydrogen Safety (ICHS 2015), Yokohama, J, October 19-21, 2015 CD-ROM Paper ID 148.

[40] Yudina, T., Filippov, A., Kiselev, A., Bleyer, A., Bentaib, A., Malet, J., Andreani, M., Paladino, D., Guentay, S., Gehr, R., Kelm, S., Allelein, H., J., Benz, S., Jordan, T., Liang, Z., Popova, T. (2015), HAMIX benchmarking tests: Code-to-code comparison. International Congress on Advances in Nuclear Power Plants (ICAPP 2015), Nice, F, May 3-6, 2015 Proceedings on USB-Stick Paper 15363 Madison, Wis.: Omnipress, 2015.

Group: AREVA Nuclear Professional School

Stochastic Field Method

Martin Raquet, Andreas Class

Introduction

If a physical process shows a spectrum of solutions whereas the engineer is only interested in mean values statistical methods are high promising methods. As an example this is the case in the field of nuclear energy regarding the neutron physics describing nuclear reactions. Furthermore turbulent premixed combustion is the area where stochastic methods have been developed. This article considers the Stochastic Field Method that belongs to the class of Monte Carlo Methods. It shows high accuracy when describing statistical processes and can be applied to numerical calculations of numerous technical problems. The basic idea behind this method is to disturb the property of a field by white noise to represent a small number of realizations of this field. Studying the evolution, i.e. multiple sets of such realizations, ensures fast statistical convergence of the method.

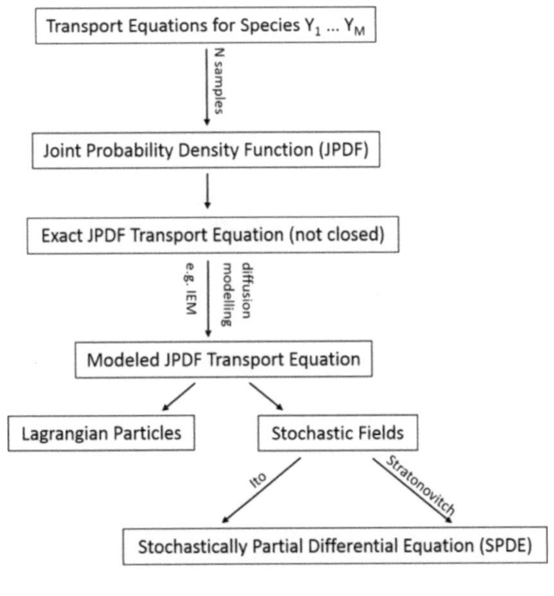

Fig. 1: Procedure when solving PDF equations for highly non-linear processes

Basics of the Stochastic Field Method

The problem of turbulent reacting flows can be boiled down to determine an accurate value of the mean chemical reaction rates. Turbulent reacting flows involve complex nonlinear reactions and a huge number of different species. These reactions unfortunately are strongly non-linear so that computing the mean reaction rate from mean species concentrations yields nonphysical data which differs from results computed from instantaneous concentrations by orders of magnitude. In order to obtain an accurate value of the mean reaction values the evolution of time resolved reaction rates could be recorded and averaged but this unfortunately requires a direct numerical simulation (DNS). An attractive alternative is to determine the joint probability density function (JPDF) of concentrations, i.e. the distribution function showing the probability of observing any combination of species concentrations. This allows computing the mean reaction rates without prior application of any averaging. The general procedure described below is visualized in Fig. 1. The establishment of an accurate JPDF is a challenge. For some well-known processes its shape can be presumed as e.g. Gaussian. If the shape is unknown or too complicated (e.g. multimodal) to be estimated by a common distribution the PDF must be created by producing samples of the M species transport equations. Later on sampling will consist of N samples (to be distinguished from the M species).

Fig. 2 shows a small number of samples (blue) clustered near the most probable property value. Sampling a large number of such samples eventually leads to the distribution (green) and the according PDF (yellow). One can imagine the PDF consisting of a huge number of delta-functions. The PDF indicates the probability for a certain concentration. The joint probability density function (JPDF) describes the probability for species combinations. Moreover the PDF and the JPDF are dependent on spatial location and on time. The physics about the transport of species is given by the species transport equation. Using this information the dependence of the PDF (JPDF respectively) on space and time can be predicted. More abstractly spoken one regards the PDF as being transported. The PDF transport equation (JPDF respectively) can be derived from the species transport equations. The mathematical procedure has been shown in (Pope S. B., 2000). The resulting PDF transport equation represents an exact formulation.

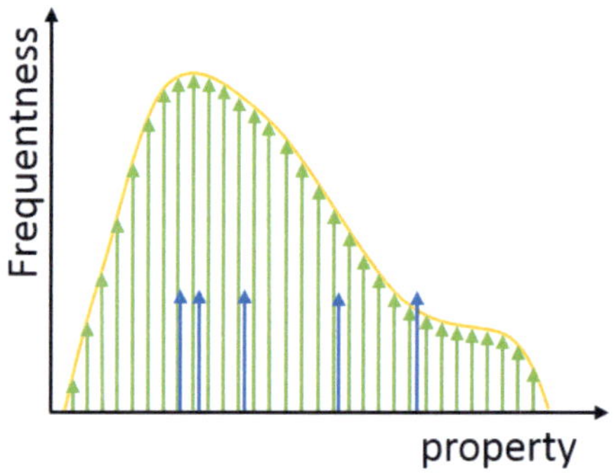

Fig. 2: The Probability Density Function of a property (yellow) is estimated by time averaging samples (blue) to obtain the corresponding distribution (green).

Unfortunately the mixing term is not closed which impels the use of models. The establishment of an adequate diffusion model has been subject to investigations during the last decades (Correa, 1995).

Several models are introduced and compared in (Edward S. Richardson, 2012). The Interaction-by-Exchange-with-the-Mean (IEM) model is a simple and common approach. The application of such a model results in the modeled PDF transport equation. Yet it is too complex to be solved directly. Instead the transport of N samples representing the PDF is investigated and ensemble averaged (compare Fig. 2). The numerical calculation of these samples of the PDF transport equation can be realized at low numerical cost. This strategy is referred to as “Monte Carlo Method”. These samples can either be regarded as (i) particles or as (ii) stochastic fields, respectively. The first method (i) uses injection of artificial particles, so called Lagrangian particles. They are observed from the Lagrangian point of view whereas the continuous phase is calculated using a Eulerian perspective. The procedure for the application on combustion is described in detail in (Pope S. B., 1980). Moreover it is shown that the description of the PDF consisting of a large number of samples (in the limit of infinite N) converges to the correct solution of the PDF transport equation. The second method (ii) treats the samples as stochastic fields. Both the samples and the flow are viewed applying a Eulerian perspective. The method was first introduced in (Valiño, 1998).

Using the method of stochastic fields the modelled PDF transport equation has to be transferred to a so called stochastically partial differential equation (SPDE) that describes the behavior of the stochastic fields (Vladimir Sabel'nikov, 2005).

Stochastic processes employ integral descriptions where two interpretations are given by Ito and Stratonocitch calculus (Peter E. Kloeden, 1999).

Valiño established his work based on (Gardiner, 1985) where an Ito interpretation was used. Alternatively in (Vladimir Sabel'nikov, 2005) a Stratonovitch interpretation was used yielding the same SPDE as previously presented in (Valiño, 1998).

Note that in Fig. 1 Lagrangian Particles on the left branch and Stochastic Fields on the right branch could be linked directly. In (O. Soulard, 2006) it is shown how to transform equations describing Lagrangian particles to equations describing Stochastic Fields (SPDEs).

Applications of the Stochastic Field Method

Originally the Stochastic Field Method was applied on problems in the field of combustion by W. P. Jones (e.g. (W. P. Jones, 2013)). Recently it has also been transferred to applications in other fields. The application on two-phase flows ((J. Dumond, 2013)) was pioneered at IKET. Similarly a generalization to chemically reacting fluids or solids obviously is possible. In this context the simulation of a bubble column reactor is envisioned. Furthermore the Stochastic Field Method could be applied on active fluids such as bacteria in micro flows.

Application on two phase flows

Cavitating flows are observed during the injection into a combustion engine. Relying on cavitation in (J. Dumond, 2013) a specific safety relevant component of GEN IV nuclear power plants was developed and investigated. The prediction of the probability to be issued with a bubble of a certain size is necessary to ensure the function of these systems. The Stochastic Field Method combined with the technique "binning" allows to visualize the corresponding PDF.

Advantages of the Stochastic Field Method

Traditionally the bubbles in a two phase flow are calculated using a Euler-Lagrange solver. Trajectories of three exemplary bubbles on their way through the discretized flow field are depicted in Fig. 3.

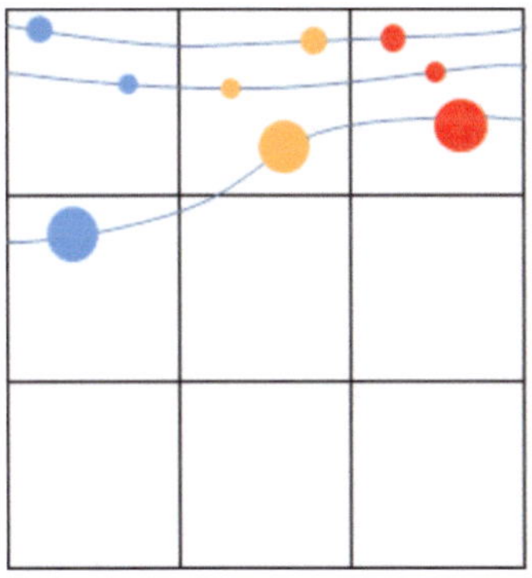

Fig. 3: Under the Lagrangian perspective single bubbles are observed on their way through the flow field. The volume fraction of the gas phase is evaluated as the mean of a representative number of bubbles. Three bubbles with their according trajectories are shown during the three time steps t1 , t2 and t3.

The usage of two different perspectives needs a coupling of fundamentally different descriptions which turns out to be CPU-intensive. Using a Euler-Euler solver the calculation is more efficient since it employs pure Eulerian description. Often the advantage of a higher efficiency is corrupted by a reduction of accuracy. The Stochastic Field Method combines advantages of both worlds: the efficiency of the Euler-Euler solver and the accuracy of the Euler-Lagrange solver. The pure Eulerian perspective eliminates the necessity of coupling and significantly decreases the calculation time and code complexity.

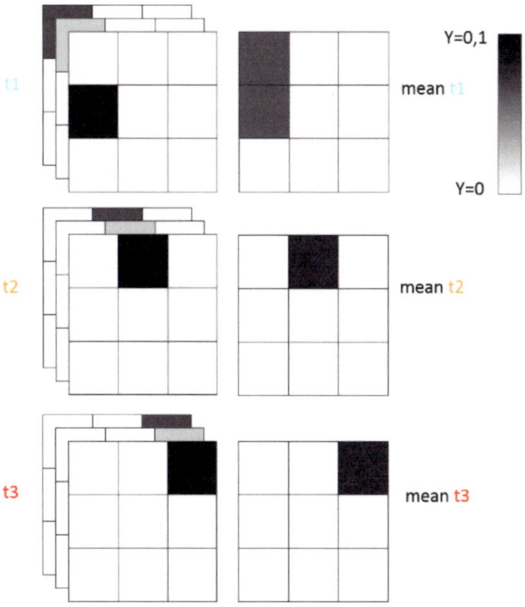

Fig. 4: Under a Euler-Euler perspective the gas and liquid phases are represented using interacting fields. The Stochastic Field Method extends this approach by creating statistical variations by disturbing the fields of the gas phase. Instead of considering Lagrangian trajectories the Stochastic Field Method takes into account statistical fluctuations. The figures shows schematically several realizations of the gas phase's volume fraction Y and the mean values during three time steps t1 , t2 and t3.

Yet more important is the fact that in the pure Eulerian approach Lagrangian particles are absent so that techniques necessary to control Lagrangian particle number density can be omitted altogether. To be statistically reliable a Euler-Lagrange-Solver needs a certain number of Lagrangian particles present in each cell. In case the number is too small the overall number of particles is to be increased or the technique of cloning has to be applied carefully. In case too many particles are present pairs can be fused to form fewer representative particles.

Instead of particles the Stochastic Field Method uses different realizations of a property each being distinct due to the influence of a randomly generated white noise. Most important is the fact that in every cell the same number of representations is generated. Thus techniques like cloning or fusion of particles are obsolete in the Stochastic Field Method which leads to a higher efficiency. A further advantage becomes apparent when considering the lifetime of fluctuations. In the Stochastic Field Method this is manually selected e.g. typically the order of the turbulent timescale. Euler-Lagrange solvers regard a particle during its complete way though the flow regime. Consequently the Stochastic Field Method provides more realizations of the stochastic process per time. Fig. 4 schematically shows how the same process of Fig. 3 would be regarded by the Stochastic Field Method. Here, correspondence of three stochastic fields and three Lagrangian bubbles is highlighted. Note that instead of bubble sizes the Stochastic Field Method uses the mass fraction or the volume fraction Y.

Current research topics

The improvement of the Stochastic Field Method for an application on two phase flows is an objective of the actual research of the ANPS group at IKET. We also like to refer to related work at the hydrogen group of IKET and specifically the ongoing doctoral thesis of M. Ranft applying the Stochastic Field Method to bubbly flow in a high pressure electrolyte.

When describing bubbles in a flow field using the Stochastic Field Method proper implementation of effects like coalescence or breakup of bubbles has to be developed. Many more physical effects including phase transfer or the lift-lorce need attention. The modelling of the momentum transport between continuous and disperse phase as well as the connection of bubble size and rising velocity are of high interest. Instead of developing new models the goal is to exploit models from existing Euler-Lagrange solvers. Focus of our investigations is to elaborate to which extent existing models may be adopted to the Stochastic Field Method.

References

[1] Correa, S. M. (1995). A Direct Comparison of Pair-Exchange and IEM Models in Premixed Combustion. *combustion and flame*, pp. 194-206.

[2] Edward S. Richardson, J. H. (2012, 03 23). Application of PDF mixing models to premixed flames with differential diffusion. *Combustion and Flame, doi:10.1016/j.combustflame.2012.02.026.*

[3] Gardiner, C. W. (1985). *Handbook of Stochastic Methods.* Berlin: Springer-Verlag.

[4] Pope, S. B. (1980, 09 25). A Monte Carlo Method for the PDF Equations of Turbulent Reactive Flow. *Combustion Science and Techology*, pp. 159-174.

[5] Pope, S. B. (2000). *Turbulent Flows.* Cambridge University Press.

[6] Valiño, L. (1998, 10 09). A Field Monte Carlo Formulation for Calculating the Probability Density Function of a Single Scalar in a Turbulent Flow. *Flow, Turbulence and Combustion*, pp. 157-172.

[7] Sabel'nikov, V., O. S. (2005, 07 06). Rapidly decorrelating velocity-field model as a tool for solving one-point Fokker-Planck equations for probability density functions of turbulent reactive scalars. *Physical Review E 72.*

Read more:

[8] Bandini, G.; Polidori, M.; Gerschenfeld, A.; Pialla, D.; Li, S.; Ma, W. M.; Kudinov, P.; Jeltsov, M.; Kööp, K.; Huber, K.; Cheng, X.; Bruzzese, C.; Class, A. G.; Prill, D. P.; Papukchiev, A.; Geffray, C.; Macian-Juan, R.; Maas, L. (2015). Assessment of systems codes and their coupling with CFD codes in thermal-hydraulic applications to innovative reactors. Nuclear Engineering and Design, 281, 22-38. doi:10.1016/j.nucengdes.2014.11.003

[9] Batta, A.; Class, A. (2015). Analysis of pressure drop in rod bundles in heavy liquid metal. 16th International Topical Meeting in Nuclear Reactor Thermal Hydraulics (NURETH-16), Chicago, Ill, August 30 - September 4, 2015.

[10] Batta, A.; Class, A. (2015). Analysis of pressure drop in rod bundles in heavy liquid metal.

16th Int. Topical Meeting in Nuclear Reactor Thermal Hydraulics (NURETH-16), Chicago, Ill, August 30 - September 4, 2015 Proceedings, 2762-2773, ANS, LaGrange Park, Ill.

[11] Batta, A.; Class, A. (2015). Thermal hydraulic numerical investigation of the heavy liquid metal free surface of MYRRHA spallation target experimental. Technology and Components of Accelerator-Driven Systems 2nd International Workshop Proceedings, Nantes, F, May 21-24, 2013, 277-282, OECD/NEA, Paris, F.

[12] Batta, A.; Class, A. G.; Litfin, K.; Wetzel, T.; Moreau, V.; Massidda, L.; Thomas, S.; Lakehal, D.; Angeli, D.; Losi, G.; Mooney, K. G.; Van Tichelen, K. (2015). Experimental and numerical investigation of liquid-metal free-surface flows in spallation targets. Nuclear Engineering and Design, 290, 107-118. doi:10.1016/j.nucengdes.2014.11.009

[13] Benz, M.; Schulenberg, T. (2015). A novel approach for turbulence modeling of wavy stratified two-phase flow. 16th International Topical Meeting in Nuclear Reactor Thermal Hydraulics (NURETH-16), Chicago, Ill, August 30 - September 4, 2015.

[14] Benz, M.; Schulenberg, T. (2015). Validation analyses of advanced turbulence model approaches for stratified two-phase flows. 8th International Conference on Computational and Experimental Methods in Multiphase and Complex Flow, Valencia, E, April 20-22, 2015.

[15] Benz, M.; Schulenberg, T. (2015). A novel approach for turbulence modeling of wavy stratified two-phase flow. 16th International Topical Meeting in Nuclear Reactor Thermal Hydraulics (NURETH-16), Chicago, Ill, August 30 - September 4, 2015 Proceedings, 1992-2005, ANS, LaGrange Park, Ill.

[16] Cheng, X.; Batta, A.; Bandini, G.; Roelofs, F.; van Tichelen, K.; Gerschenfeld, A.; Prasser, M.; Papukchiev, A.; Hampel, U.; Ma, W. M. (2015). European activities on crosscutting thermal-hydraulic phenomena for innovative nuclear systems. Nuclear Engineering and Design, 290, 2-12. doi:10.1016/j.nucengdes.2014.11.007

[17] Class, A. G.; Viellieber, M.; Bruzzese, C.; Raquet, M.; Dumond, J. (2015). Thermohydraulic codes applied to wind power and combustion engines. 46.Jahrestagung Kerntechnik, Berlin, 5.-7.Mai 2015.

[18] Fetzer, J. R.; Class, A. G.; Fazio, C.; Gordeev, S. (2015). Conceptual design studies for the liquid metal spallation target META:LIC. Technology and Components of Accelerator-Driven Systems. 2nd International Workshop Proceedings, Nantes, F, May 21-24, 2013 Paris, F, 309-319, OECD/NEA, Paris, F.

[19] Hartig, M.; Schulenberg, T.; Batta, A. (2016). Numerical study of fluid-structure interaction with heat transfer at supercritical pressure in a fuel rod assembly. Journal of Nuclear Engineering and Radiation Science, 2, 011013/1-12. doi:10.1115/1.4031816

[20] Hartig, M.; Batta, A.; Schulenberg, T. (2015). Numerical study of fluid-structure interaction with heat transfer at supercritical pressure in a fuel rod assembly. 7th International Symposium on Supercritical Water-Cooled Reactors (ISSCWR 2015), Helsinki, SF, March 15-18, 2015.

[21] Hartig, M.; Batta, A.; Schulenberg, T. (2015). Numerical study of fluid-structure interaction with heat transfer at supercritical pressure in a fuel rod assembly. Proceedings of the 7th International Symposium on Supercritical Water-Cooled Reactors (ISSCWR 2015), Helsinki, SF, March 15-18, 2015. Ed.: S. Penttilä, Paper ISSCWR7-20137, VTT, Espoo.

[22] Heinze, D. (2015). Physically-Based Models for Two-Phase Flow Phenomena in Steam Injectors: A One-Dimensional Simulation Approach (KIT Scientific Reports ; 7704). Dissertation. KIT Scientific Publishing, Karlsruhe. doi:10.5445/KSP/1000048586

[23] Heinze, D.; Schulenberg, T.; Behnke, L. (2015). A physically based, one-dimensional two-fluid model for direct contact condensation of steam jets submerged in subcooled water. Journal of Nuclear Engineering and Radiation Science, 1 (2), 021002/1-8. doi:10.1115/1.4029417

[24] Hornung, C.; Class, A. G.; Viellieber, M. (2015). Simulation of wake effects in windfarms using an actuator disk. GAMM 86th Annual

Scientific Conference, Leece, I, March 23-27, 2015.

[25] Hornung, C.; Viellieber, M.; Class, A. (2015). Simulation of wake effects in windfarms using an actuator disk implementation. Proceedings in Applied Mathematics and Mechanics, 15, 487-488. doi:10.1002/pamm.201510234

[26] Koopman, H. K. (2015). Analytical investigations concerning the performance of vane separators and experimental validation of droplet separation efficiency (KIT Scientific Reports ; 7690). Dissertation. KIT Scientific Publishing, Karlsruhe. doi:10.5445/KSP/1000045141

[27] Litfin, K.; Fetzer, J. R.; Batta, A.; Class, A. G.; Wetzel, T. (2015). High power spallation target using a heavy liquid metal free surface flow. Journal of Radioanalytical and Nuclear Chemistry, 305, 795-803. doi:10.1007/s10967-015-4002-z

[28] Noack, R.; Class, A. G. (2015). Coarse-Grid-CFD method applied to two-phase flow problems. 46.Jahrestagung Kerntechnik, Berlin, 5.-7.Mai 2015, CD-ROM, INFORUM GmbH, Berlin.

[29] Pacio, J.; Litfin, K.; Batta, A.; Viellieber, M.; Class, A.; Doolaard, H.; Roelofs, F.; Manservisi, S.; Menghini, F.; Böttcher, M. (2015). Heat transfer to liquid metals in a hexagonal rod bundle with grid spacers: Experimental and simulation results. Nuclear Engineering and Design, 290, 27-39. doi:10.1016/j.nucengdes.2014.11.001

[30] Raque, M.; Schulenberg, T.; Zeiger, T. (2016). Expected safety performance of the supercritical water reactor fuel qualification test. Journal of Nuclear Engineering and Radiation Science, 2, 011005/1-9. doi:10.1115/1.4030917

[31] Rohde, M.; Peeters, J. W. R.; Pucciarelli, A.; Kiss, A.; Rao, Y. F.; Onder, E. N.; Muehlbauer, P.; Batta, A.; Hartig, M.; Chatoorgoon, V.; Thiele, R.; Chang, D.; Tavoularis, S.; Novog, D.; McClure, D.; Gradecka, M.; Takase, K. (2016). A blind, numerical benchmark study on supercritical water heat transfer experiments in a 7-rod bundle. Journal of Nuclear Engineering and Radiation Science, 2 (2), 021012. doi:10.1115/1.4031949

[32] Rohde, M.; Peeters, J. W. R.; Pucciarelli, A.; Kiss, A.; Rao, Y. F.; Onder, E. N.; Mühlbauer, P.; Batta, A.; Hartig, M.; Chatoorgoon, V.; Thiele, R.; Chang, D.; Tavoularis, S.; Novog, D.; McClure, D.; Gradecka, M.; Takase, K. (2015). A blind, numerical benchmark study on supercritical water heat transfer experiments in a 7-rod bundle. Proceedings of the 7th International Symposium on Supercritical Water-Cooled Reactors (ISSCWR 2015), Helsinki, SF, March 15-18, 2015. Ed.: S. Penttilä, VTT, Espoo.

[33] Ruzickova, M.; Vojacek, A.; Schulenberg, T.; Visser, D. C.; Novotny, R.; Kiss, A.; Maraczy, C.; Toivonen, A. (2016). European Project 'Supercritical Water Reactor-Fuel Qualification Test': Overview, Results, Lessons Learned, and Future Outlook. Journal of Nuclear Engineering and Radiation Science, 2, 011002/1-10. doi:10.1115/1.4031034

[34] Ruzickova, M.; Vojacek, A.; Schulenberg, T.; Visser, D. C.; Novotny, R.; Kiss, A.; Maraczy, C.; Toivonen, A. (2015). European project 'Supercritical Water Reactor - Fuel Qualification Test' (SCWR-FQT): Overview, results, lessons learnt and future outlook. Proceedings of the 7th International Symposium on Supercritical Water-Cooled Reactors (ISSCWR 2015), Helsinki, SF, March 15-18, 2015. S. Penttilä, Penttilä, S. [Hrsg.] Proceedings of the 7th International Symposium on Supercritical Water-Cooled Reactors (ISSCWR 2015), Helsinki, SF, March 15-18, 2015 Espoo: VTT, 2015 Paper No. ISSCWR7-2054 (VTT Technology ; 216), PaperNo.ISSCWR7-2054, VTT, Espoo.

[35] Schulenberg, T. (2015). Heat transfer during depressurization of supercritical steam. EST 2015 - Energy, Science and Technology, Karlsruhe, May 20-22, 2015 Book of Abstracts.

[36] Schulenberg, T. (2016). Guest Editorial - Special Issue: Seventh International Symposium on Supercritical Water-Cooled Reactors. Journal of Nuclear Engineering and Radiation Science, 2, 010301/1-2. doi:10.1115/1.4031661

[37] Schulenberg, T.; Raque, M.; Zeiger, T. (2015). Expected safety performance of the SCWR fuel qualification test. Proceedings of the 7th International Symposium on Supercritical Water-Cooled Reactors (ISSCWR 2015),

Helsinki, SF, March 15-18, 2015. Ed.: S. Penttilä, PaperNo.ISSCWR7-2033, VTT, Espoo.

[38] Viellieber, M. O.; Class, A. G. (2015). Coarse-grid-CFD for the thermal hydraulic investigation of rod-bundles. Proceedings in Applied Mathematics and Mechanics, 15, 497-498. doi:10.1002/pamm.201510239

[39] Viellieber, M. O.; Class, A. G. (2015). Coarse-grid-CFD for the thermal hydraulic investigation of Rod-bundles. GAMM 86th Annual Scientific Conference, Leece, I, March 23-27, 2015.

[40] Vojacek, A.; Ruzickova, M.; Schulenberg, T. (2016). Design of an in-pile SCWR fuel qualification loop. Journal of Nuclear Engineering and Radiation Science, 2, 011003/1-7. doi:10.1115/1.4030872

[41] Vojacek, A.; Ruzickova, M.; Schulenberg, T. (2015). Design of an in-pile SCWR fuel qualification test loop. Proceedings of the 7th International Symposium on Supercritical Water-Cooled Reactors (ISSCWR 2015), Helsinki, SF, March 15-18, 2015. Ed.: S. Penttilä, PaperNo.ISSCWR7-2065, VTT, Espoo.

[42] Weisenburger, A.; Müller, G.; Bruzzese, C.; Class, A. G. (2015). Coolant chemistry control. Oxygen mass transport in lead bismuth eutectic. Actinide and Fission Product Partitioning and Transmutation: 13th Information Exchange Meeting, Seoul, Korea, September 23-26, 2014 Proceedings, 211-219, OECD/NEA, Paris, F.

[43] Zeiger, T. (2015). Auswirkungen eines postulierten Druckrohrversagens während des SCWR Fuel Qualification Tests (KIT Scientific Reports; 7687). Dissertation. KIT Scientific Publishing, Karlsruhe. doi:10.5445/KSP/1000044861

Experimental investigation on the physical and chemical properties of geothermal brine under in-situ conditions

S. Herfurth, P. Orywall

The group Energy and Process Engineering contributes to the field of efficient use of geothermal energy within the PoF programme Renewable Energies. In 2015 a new project was established to determine physical properties of geothermal brine as a result of a long term development. The following chapter describes the scope of this project.

Precipitation and scaling in the thermal water loop leads to damage of components and blockage of pipes in the loop. To increase the availability of geothermal power plants it is necessary to acquire a detailed understanding of the chemical processes in the brine which lead to supersaturation. This affects not only the surface installations but also the conditions and reactions in the reservoir after reinjection. The HydRA-test facility allows for experiments to investigate fluid-rock interaction including precipitation. The set-up as well as first results of the experiments are described below.

PEther – Physical Properties of Thermal Water under In-situ-Conditions

PEther is a research project funded by the German Federal Ministry for Economic Affairs and Energy (BMWi, reference number: 0325761). The project partners are KIT (Karlsruhe Institute of Technology; technical project coordination), GeoT (GeoThermal Engineering GmbH, Karlsruhe; geoscientific project coordination), gec-co (Global Engineering and Consulting GmbH, Augsburg) and as subcontractor GTN (Geothermie Neubrandenburg GmbH).

The objective of PEther is to determine thermo-physical properties (specific isobaric heat capacity, kinematic viscosity, density and thermal conductivity) of geothermal brine under conditions that reflect the original conditions to be found in geothermal applications (pressure, temperature, chemical composition including gas content of the brine). The knowledge of these thermo-physical properties is crucial in order to estimate the thermal output and therefore the economic viability as well as to optimize the plant or heating station design during the development and operation of a geothermal project. Up to now, only a limited number of measurements of selected physical properties have been made, usually under laboratory conditions as well as for individual geothermal plants. Currently available mathematical formulae describing the thermo-physical properties are typically not valid for the conditions to be found in geothermal applications and do not consider the substantial influence of the chemical composition of the thermal water. Also, actual geothermal brines have not been subject of systematic measurements performed under operational conditions on a large-scale basis.

In this project, the thermo-physical properties of geothermal brine are measured under in-situ-conditions as functions of pressure, temperature and salinity. The measurements are taking place directly at several geothermal applications located in Germany's hydro-geothermal key regions (see Fig. 1) and in the laboratory. Locations for measurements are selected upon geological criteria. Until now five in-situ measurements have been successfully performed: Two of them in the Upper Rhine Valley and three in the Molasse Bassin. Furthermore, seven in-situ campaigns are planned and comprehensive test series with artificial brine (mixtures of desalinated water and salts: $NaCl$, $CaCl_2$, KCl, $MgCl_2$, Na_2SO_4, $CaSO_4$ and $NaHCO_3$) are performed in the lab (still on-going). The results will then be compared to the in-situ measurements and serve for model development and software validation.

Fig. 1: Map of the geothermal regions in Germany

To meet any geothermal reservoir conditions to be found in Germany, a mobile testing unit was developed at the institute with instruments specifically designed in-house. This testing unit consists of four Measurement devices (see Fig. 2 & 3): a flow calorimeter (cp), a viscosimeter (v), a densimeter (ρ) and a thermal conductivity meter (λ). The devices for v, cp and ρ have been developed and tested at IKET

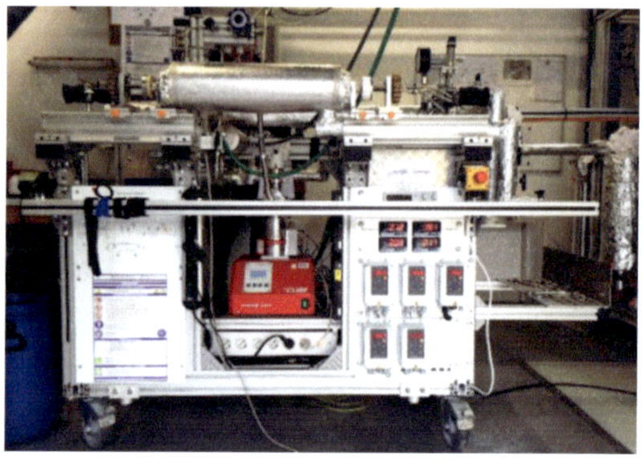

Fig. 2: Mobile testing unit for the investigation of physical fluid properties under reservoir conditions

They are designed for a temperature range of 20 °C to 170 °C, pressure up to 30 bar and need to withstand salinities in the range of 0 to 360 g/l. The λ-device is commercial (Lambda 1, F5 Technology) and just slightly modified regarding the filling. Details on the measurement devices can be found in Schröder et al. (2015, [9]).

Pressure and temperature are controlled and kept constant in the testing unit. The whole unit weighs ~250 kg and its dimensions are: 1.90 m length x 1.25 m width x 1.60 m height. Due to the compact design it is possible to move the unit easily to every geothermal location.

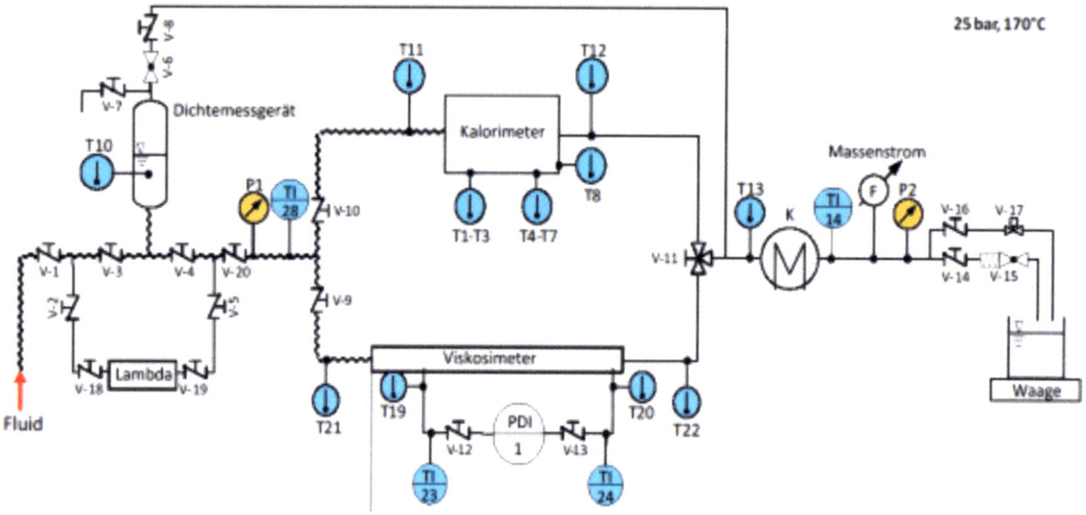

Fig. 3: Process and instrumentation diagram of the mobile testing unit. The thermal water flows from left to right. The measurement devices are partly built in in series, partly parallel, but they are operated separately while measuring.

All measuring devices have been tested with pure water in order to compare the recorded data to the literature: The variation for the heat capacity cp amounts to 1% (literature: Lemmon et al (2010)). According to Picker et al. (1971), Perron et al. (1975) and Desnoyers et al. (1976), who also did experiments with flow calorimeters, the variation is within the uncertainty of measurement. Apart from pure water, aqueous NaCl-solutions have been measured and the results compared to literature (Thurmond & Brass, 1988, Smith-Magowan & Wood, 1981, Hnědkovský et al., 2002, Picker et al., 1971, Perron et al., 1975 und Desnoyers et al., 1976); it results in a variation of 1%. The data for kinematic viscosity ν was checked against literature data taken from NIST data base (Lemmon et al., 2010) and Kestin et al. (1981b). Here, the variation lies within 3% for pure water and aqueous solutions. For determining the density ρ, a weighing method is used. Therefore, the volume of the container in use is measured accurately (~0.1% accuracy) and the tare weight is determined with 0.4% accuracy. The obtained data has been compared to Lemmon et al. (2010) and a variation of 0.4% was calculated (only for pure water). The comparison is shown in Herfurth et al. (2015) in all detail [2], [3].

First results of in-situ measurements and lab measurements with the particular geothermal brine are presented in Fig. 4. This figure shows the physical properties of geothermal brine under in situ conditions at five different geothermal sites in Germany. The top left diagram shows the heat capacity, the top right one shows density, the bottom left is for kinematic viscosity and the bottom right for heat conductivity. The different geothermal sites operate at different thermal water temperature and pressure. The effect of pressure seems to be negligible; the influence of temperature follows the expected trends. Over all, the data shows a strong dependency on salinity. Data from lab measurements with artificial brine are not yet available.

The results from in-situ measurements, lab measurements with geothermal and artificial brine will be compared to standard analytical methods as well as used to calibrate laboratory

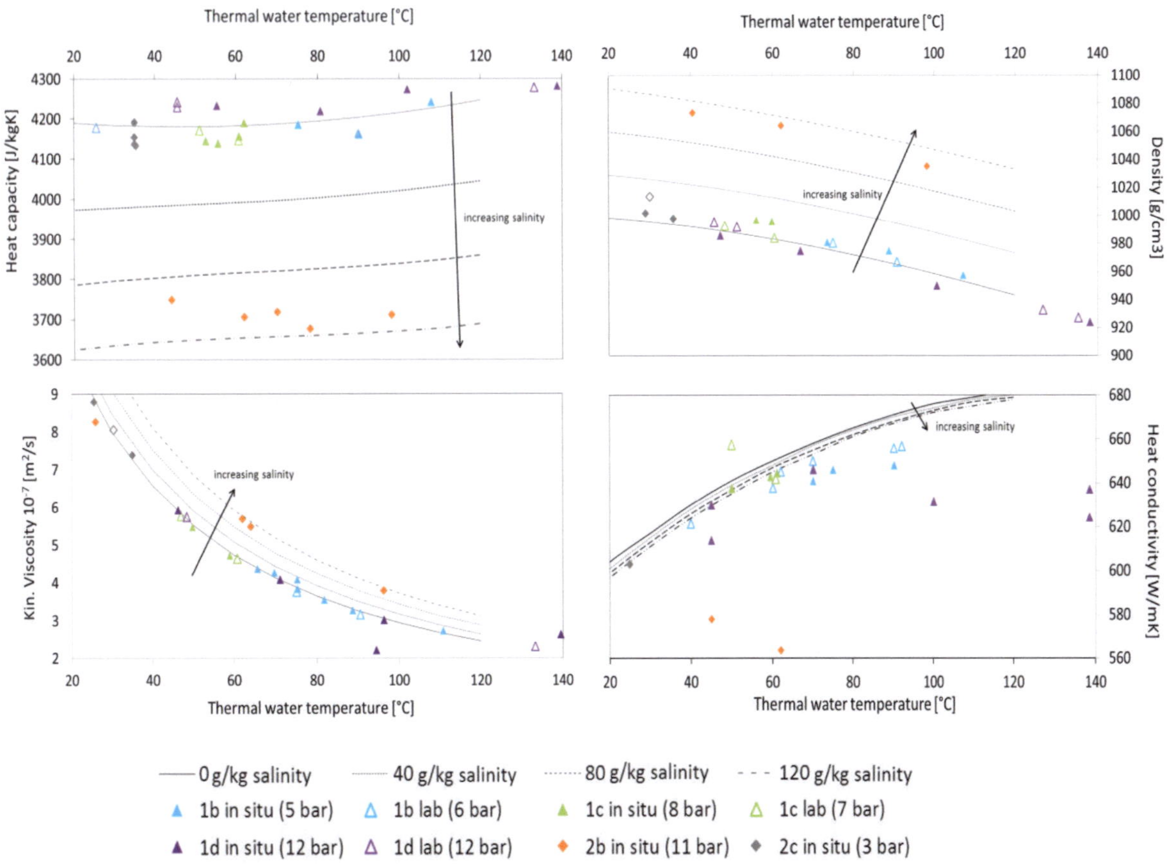

Fig. 4: Physical properties – heat capacity, density, kinematic viscosity and heat conductivity – of thermal water from different locations over thermal water temperature (see Herfurth et al. (2016). Curves (dotted, dashed or line) are taken from seawater correlations (Sharqawi et al., 2010) as orientation concerning salinity.

measurements that simulate the encountered in-situ conditions. A comprehensive series of measurements is performed in order to create a Germany-wide data base and to model the thermal water's thermo-physical characteristics as a basis for future designs of geothermal applications.

HydRA -Hydrothermal Reaction Apparatus

Induced barite scaling in porous geothermal reservoir rocks during core-flooding experiments

Barite scaling that leads to reservoir-clogging is a well-studied (threat) scenario in oil exploration when sulfate-containing seawater is injected into the reservoir hosting barium-rich formation water (Todd & Yuan, 1990). In geothermal exploration barite supersaturation and precipitation is achieved by extracting the heat from a reservoir fluid which is saturated or slightly undersaturated regarding barite (He et al., 1994). Either barite particles nucleate from the spent fluid and clog the pore throats or precipitation occurs in the available pore space by ongrowth on matrix particles. One good example is the case of Soultz-sous-Forêts (France) where fluid circulation leads to massive deposits in the tube on the reinjection side, whereas the downstream fate of barite is virtually unknown (Scheiber et al., 2014).

This experimental research focuses on the near-wellbore reinjection environment which is simulated in the laboratory with a core-flooding apparatus shown in Fig. 5. In further steps the coupling between thermo-chemical hydraulic processes is of particular interest and will be complemented by reactive transport modelling using a well-known open source code algorithm (Orywall et al., 2015).

In the first experimental runs a defined supersaturation with respect to barite (saturation index calculated with PhreeqC (Parkhurst & Appelo, 2013)), is achieved by mixing sodium sulfate solution (Na_2SO_4) with barium chloride solution ($BaCl_2$) shortly before the inflow of the rock sample (as porous media it is used a sandstone). As a background electrolyte an artificial fluid is mixed with regard to the natural fluid of Soultz-sous-Forêts (Mundhenk et al., 2013). The supersaturation regarding barite ($BaSO_4$) is on top of this artificial fluid. The experimental parameters like temperature and pressure are closely aligned to the operational parameters found at geothermal sites in the Upper Rhine Valley, for instance Soultz-sous-Forêts (France) and Bruchsal (Germany) (Herzberger et al., 2010; Orywall et al., 2009).

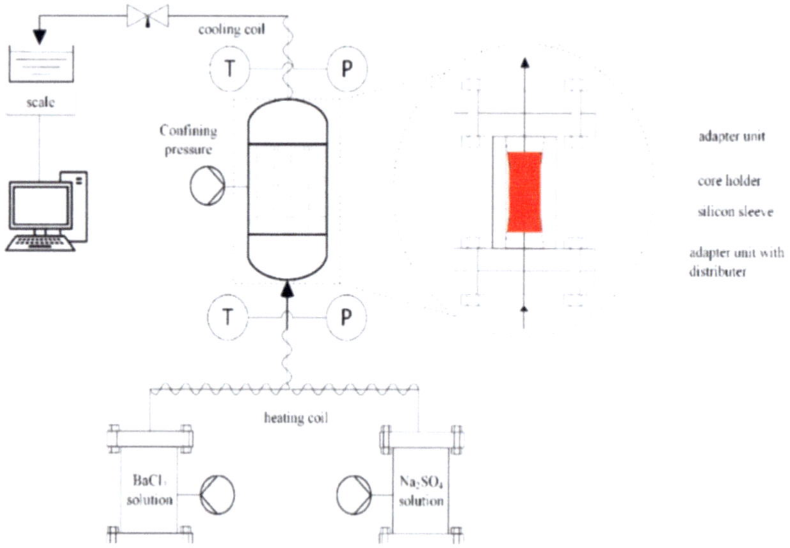

Fig. 5: The scheme of the experimental setup HydRA. In the two reservoir tanks are the separated solutions, each of them were heated to the required temperature (heating coil) and shortly before the rock sample (built in the autoclave tight enclosed by the red tube) the two solutions get mixed and the supersaturated solution flow through the porous rock sample. For the collection of the fluid samples, the solution was cooled down (cooling coil). The whole system is monitored by some pressure and temperature sensors and all generated data were digitally recorded by the Software OPAL (Daubner & Krieger, 2010).

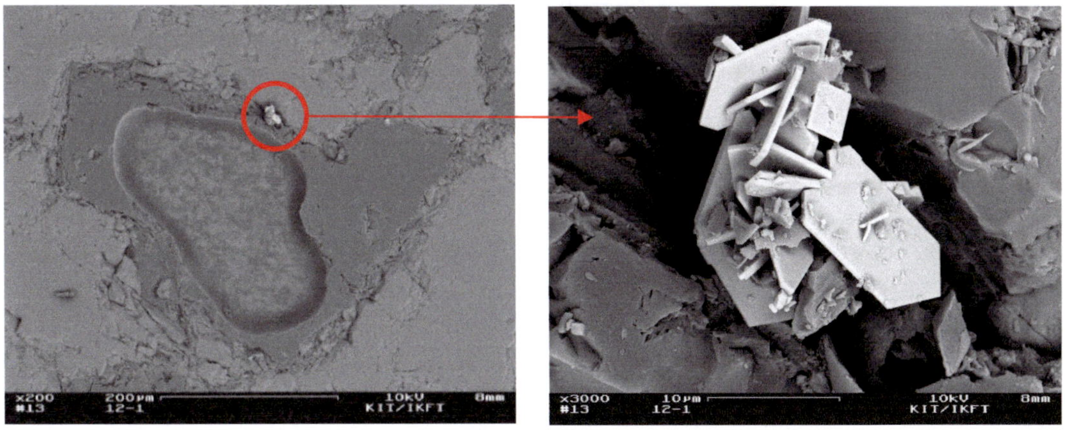

Fig. 6a) und b) – barite crystal in the pore space of the rock sample

First analytical results of the rock samples by using SEM (scanning electron microscopy) show expected barite crystals in the pore space of the perfused sandstone (Fig. 6a),b)). The crystal shapes of the newly formed ones are different, depending on the location where they were found. This one was detected at the outflow of the rock sample.

Among other analytical work, the permeability of the rock samples can be measured in-situ with HydRA. The determination of this hydraulic parameter is performed before and after the core-flooding tests. With the mentioned super-saturation of barite, the expected precipitations change the permeability of the used rock sample during the experiments. First results show exactly this behavior. The change of the per-

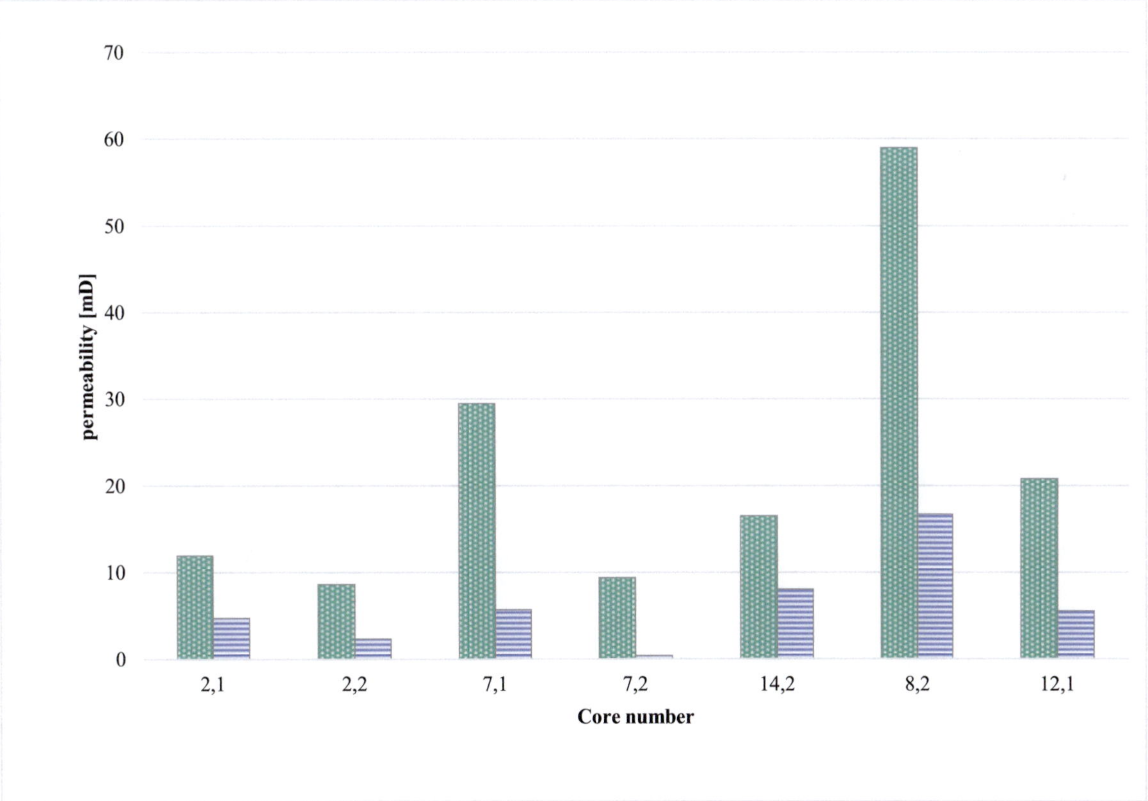

Fig. 7: The permeability is a hydraulic characteristic of natural rocks. It depends on different values like flow rate, viscosity of the fluid, differential pressure over the sample and size of the used sample. The unit "D" (Darcy) can be converted into the SI Unit by 1D = $9.86923 \cdot 10^{-13} m^2$. The x-coordinate shows the rock sample number from the different experiments. For each experiment a new rock sample were used.

meability is shown in Fig. 7. The green (dotted) bars represent the permeability before and the blue (striped) bars show the values after the core-flooding experiments.

Without exceptions, all values are higher before and lower after. This confirms the permeability decrease of the porous rock sample through precipitations into the pore spaces.

Further work should investigate the temperature and pressure dependence of the crystal formation. Additional the water-rock interactions will be studied, to answer the question whether there is only the provoked precipitation in the rock sample or additional some dissolution processes of the initial sandstone mineral phases.

References

[1] Desnoyers, J.E., Visser, C., Perron, G., Picker, P., Re-examination of the Heat Capacities Obtained by Flow Microcalorimetry. Recommendation for the Use of a Chemical Standard, *Journal of Solution Chemistry*, 5, (1976), 605-616.

[2] Herfurth, S.; Schröder, E.; Thomauske, K.; Kuhn, D. (2015). Messung physikalischer Thermalwassereigenschaften unter In-situ Bedingungen. Geothermiekongress 2015, Essen, 2.-4. November 2015.

[3] Herfurth, S., Schröder, E., Thomauske, K., Herberger, S., PETher – Physical Properties of Thermal Water under In-situ-Conditions, *Proceedings European Geosciences Union - General Assembly 2016,* Vienna (2016).

[4] Hnedkovsky, L., Hynek, V., Majer, V., Wood, R.H., A New Version of Differential Flow Heat Capacity Calorimeter; Tests of Heat Loss Corrections and Heat Capacities of Aqueous NaCl from T = 300 K to T = 623 K, *The Journal of Chemical Thermodynamics*, 34, (2002), 755-782.

[5] Kestin, J., Khalifa, H.E., Correia, R.J., Tables of the Dynamic and Kinematic Viscosity of Aqueous NaCl Solutions in the Temperature Range 20-150 °C and the Pressure Range 0.1-35 MPa, *Journal of Physical and Chemical Reference Data*, 10, (1981b), 71-88.

[6] Lemmon, E.W., Huber, M.L., McLinden, M.O., Nist Reference Fluid Thermodynamic and Transport Properties, Refprop. in Physical and Chemical Properties Division, National Institute of Standards and Technology, Boulder, Colorado 80305 U.S., (2010).

[7] Perron, G., Fortier, J.-L., Desnoyers, J.E., The Apparent Molar Heat Capacities and Volumes of Aqueous NaCl from 0.01 to 3 mol kg−1 in the Temperature Range 274.65 to 318.15 K, *The Journal of Chemical Thermodynamics,* 7, (1975), 1177-1184.

[8] Picker, P., Leduc, P.-A., Philip, P.R., Desnoyers, J.E., Heat Capacity of Solutions by Flow Microcalorimetry, *The Journal of Chemical Thermodynamics,* 3, (1971), 631-642.

[9] Schröder, E., Thomauske, K., Herberger, S., Schmalzbauer, J., (2015). Measuring techniques for in situ measurements of thermodynamic properties of geothermal water. World Geothermal Congress (WGC 2015), Melbourne, AUS, April 19-24, 2015 Papers publ. online Paper 15022 International Geothermal Association (IGA), 2015.

[10] Sharqawi, M.H., Lienhard V, J.H., Zubair, S.M., Thermophysical Properties of Seawater: A Review of Existing Correlations and Data, *Desalination and Water Treatment,* 16, (2010), 354-380.

[11] Smith-Magowan, D., Wood, R.H., Heat Capacity of Aqueous Sodium Chloride from 320 to 600 K Measured with a New Flow Calorimeter, *The Journal of Chemical Thermodynamics*, 13, (1981), 1047-1073.

[12] Thurmond, V.L., Brass, G.W,: Activity and Osmotic Coefficients of Sodium Chloride in Concentrated Solutions from 0 to - 40°C, *Journal of Chemical & Engineering Data*, 33, (1988), 411-414.

[13] Daubner, M., Krieger, V. (2010), Betriebsmessdatenvisualisierung und -erfassung mit OPAL (OPC-Panel Livegraph), KIT Scientific Reports 7558, Karlsruhe.

[14] He, S., Oddo, J., Tomson, M. (1994), The inhibition of gypsum and barite nucleation in NaCl brines at temperatures from 25 to 90°C, Applied Geochemistry Vol. 9, 561-567.

[15] Herzberger, P., Münch, W., Kölbel, T., Bruchmann, U., Schlagermann, P., Hötzl, H., Wolf, L., Rettenmaier, D., Steger, H., Zorn, R., Seibt, P., Möllmann, G.-U., Sauter M., Ghergut, J., Ptak T. (2010), The Geothermal Power Plant Bruchsal, Proceedings World Geothermal Congress 2010, Bali Indonesia.

[16] Mundhenk, N., Huttenloch, P., Kohl T., Steger, H., Zorn, R. (2013), Metal corrosion in geothermal brine environments of the Upper Rhine Valley–Laboratory and on-site studies, Geothermics 46, 14-21.

[17] Orywall, P.; Dimier, A.; Kohl, T.; Kuhn, D.; Place, J.; Zorn, R. (2015). Experimental validation of a numerical model. Application of an open source algorithm towards geothermal conditions. World Geothermal Congress (WGC 2015), Melbourne, AUS, April 19-24, 2015 Papers publ. online Paper 23008 International Geothermal Association (IGA), 2015 Also publ. on CD-ROM.

[18] Orywall, P., Kölbel, T., Münch, W., Genter, A., Graff, J-J., Cuenot, N., Teza, D., Hettkamp, T. (2009), Das EGS-Projekt Soultz-sous-Forêts: Von der Reservoirentwicklung zur Stromproduktion. Jahresmagazin 12 BBR Fachmagazin für Brunnen-und Leitungsbau, S. 52-59.

[19] Parkhurst, D.L., and Appelo, C.A.J. (2013): Description of input and examples for PHREEQC version 3—A computer program for speciation, batch-reaction, one-dimensional transport and inverse geochemical calculations, U.S. Geological Survey Techniques and Methods, book 6, chap. A 43, 497 p.

[20] Scheiber, J., Seibt, A., Birner, J., Cuenot, N., Genter, A., Moeckes, W. (2014), Barite scale control at the Soultz -sous- Forêts (France) EGS Site, Proceedings Workshop on Geothermal Reservoir Engineering Stanford.

[21] Todd, A.C., Yuan, M. (1990): Barium and strontium sulphate solid-solution formation in relation to North Sea scaling problems. SPE Production Engineering 5, 279-285.

Read more:

[22] Canic, T.; Baur, S.; Bergfeldt, T.; Kuhn, D. (2015), Influences on the barite precipitation from geothermal brines. World Geothermal Congress (WGC 2015), Melbourne, AUS, April 19-24, 2015 Papers publ. online Paper 27009 International Geothermal Association (IGA), 2015.

[23] Herfurth, S. (2015), Geothermie. Woher kommt sie? Wo finde ich sie? Wie kann ich sie nutzen?. Science Camp Geothermie, Karlsruhe, 1.-6. November 2015.

[24] Herfurth, S.; Gauß, M.; Henken, C. (2015), Science Camp Geothermie - Jugend trifft Wissenschaft. Geothermiekongress, Essen, 2.-4. November 2015.

[25] Herfurth, S.; Kuhn, D.; Wiemer, H. J. (2015), Performance optimization of ORC power plants. World Geothermal Congress (WGC 2015), Melbourne, AUS, April 19-24, 2015 Papers publ. online Paper 26045 International. Geothermal Association (IGA), 2015.

[26] Orywall, P.; Kuhn, D.; Mundhenk, N.; Mayer, D.; Kohl, T. (2015), Perkolation poröser Medien unter tiefengeothermischen Bedingungen. Untersuchung hydraulischer Effekte als Folge von induziertem Baryt-Scaling. Geothermiekongress 2015, Essen, 2.-4.November 2015.

[27] Schröder, E.; Kuhn, D. (2015), Geothermale Energiegewinnung. Einblicke und Funktionsweise. Geothermie-Infoveranstaltung, Seeon-Seebruck, 22. Januar 2015.

[28] Schröder, E., Thomauske, K., Gebert, C., Schmalzbauer, J.; Herberger, S., Velevska, M. (2015), Design and test of a new flow calorimeter for online detection of geothermal water heat capacity. Geothermics, 53, 202-212. doi:10.1016/j.geothermics.2014.06.001

[29] Vetter, C.; Wiemer, H. J. (2015). Dynamic simulation of a supercritical ORC using low-temperature geothermal heat. Vortr.: University of Queensland, Brisbane, AUS, 28.April 2015.

[30] Vetter, C.; Wiemer, H. J. (2015), Dynamic simulation of a supercritical ORC using low-temperature geothermal heat. World Geothermal Congress (WGC 2015), Melbourne, AUS, April 19-24, 2015 Papers publ. online Paper 26031 International Geothermal Association (IGA), 2015.